New Landscapes 景界 3

Leisure and Tourism Landscape
休闲度假景观

佳图文化 编

图书在版编目（CIP）数据

景界.3，休闲度假景观 / 佳图文化编. —广州：华南理工大学出版社，2013.8
ISBN 978-7-5623-3949-6

Ⅰ.①景…　Ⅱ.①佳…　Ⅲ.①文娱活动-景观设计-作品集-世界　②旅游区-景观设计-作品集-世界
Ⅳ.① TU986.2

中国版本图书馆 CIP 数据核字（2013）第 119065 号

景界 3：休闲度假景观
佳图文化 编

出 版 人：	韩中伟
出版发行：	华南理工大学出版社
	（广州五山华南理工大学 17 号楼，邮编 510640）
	http://www.scutpress.com.cn　E-mail: scutc13@scut.edu.cn
	营销部电话：020-87113487　87111048（传真）
策划编辑：	赖淑华
责任编辑：	张　媛　赖淑华
印 刷 者：	利丰雅高印刷（深圳）有限公司
开　　本：	1016mm×1370mm　1/16　印张：16.5
成品尺寸：	245mm×325mm
版　　次：	2013 年 8 月第 1 版　2013 年 8 月第 1 次印刷
定　　价：	296.00 元

版权所有　盗版必究　　印装差错　负责调换

Preface 前言

With the popularization of tourism as a social fashion, more and more people are fond of casual travelling. Instead of simple leisure demands, they have more stereoscopic and varied demands for leisure and tourism landscape, whose design consequently becomes more diversified for travel enjoyment. It has become the designers' focus that how to integrate into the surrounding cultural environment, how to present the local characteristics, and how to harmonize with the nature etc.

The book has selected the classical projects of leisure and tourism landscape around the world, comprehensively introducing the fascination covering tourist resorts, urban parks, ecology parks, culture parks, community parks and CBD parks etc. Regarding content arrangement, it starts the analysis from the key points, features and design points as well as a great quantity of professional drawings i.e. plane drawings, detail drawings, diagrams and photos, rich and informative.

As the latest professional Publication of New Landscape, the selected projects of this book are perfect integration of inspiration and technology, standing for the advancing design ideas of leisure and tourism landscape currently. We believe it will bring the designers and readers visual enjoyment and inspiration through the rich content, meticulous layout and fresh presentation.

随着旅行成为一种社会时尚，越来越多的人开始加入"说走就走"的旅行队伍，人们对休闲度假景观的要求也从单一的休闲需求上升为更立体化的多方面需求。为了满足人们的旅行享受，休闲度假景观的设计就变得更加多元化了。休闲度假景观的设计如何融入周遭的文化氛围、如何体现地域特色、如何与自然共生等都成为设计师们的思虑所在。

本书所选案例均为世界各地休闲度假景观的经典之作。全书分别从旅游度假公园、城市公园、生态公园、文化公园、社区公园及 CBD 公园等方面全方位地介绍休闲度假景观引人入胜之处。在内容编排上，分别从景观案例的关键点、亮点、设计要点入手，配合大量的专业技术图纸，如平面图、细部图、效果图以及实景图等，资料丰富而详实。

作为景界的最新专业读本，本书精选的案例可谓是设计灵感的迸发与技术完美结合的成果，代表了当下休闲度假景观设计的前沿理念。通过丰富的内容、精心的排版以及新颖的展示，我们相信本书定会给景观设计师及相关行业的读者带来视觉享受和设计启迪。

CONTENTS 目录

Tourism Park 旅游度假公园

- **002** Anaklia Costal Park　　Anaklia 海滨公园
- **008** Membrane Roof Construction: Palais Thermal, Bad Wildbad, Germany　　德国巴特维尔德巴德 Palais Thermal 膜结构屋顶
- **012** Wilmington Waterfront Park　　威尔明顿海滨公园
- **018** Fortress Ehrenbreitstein　　Ehrenbreitstein 堡垒
- **026** The Big Dig, Xi'an Garden Show　　西安公园展——大地洞
- **032** Trollstigen Tourist Route Project　　特洛斯蒂国家观光旅游线路

City Park 城市公园

- **046** Darling Harbour　　达令广场
- **056** Solberg Rest Station　　索伯格休息站
- **068** Observation Balloon Preview Park Orange County Great Park　　橙郡大公园之气球全景预览公园
- **074** Hengrove Leisure Centre　　Hengrove 休闲中心
- **078** Bijlmer Park　　比尔梅公园
- **086** Madrid-Río / Manzanares Lineal Park　　马德里曼萨纳雷斯线性公园
- **096** Landscape Design of Melbourne Children's Art Amusement Park　　墨尔本儿童艺术游乐园景观设计
- **102** Cutty Sark Gardens　　卡蒂萨克花园

Ecology Park 生态公园

- **110** Olive Grove, Recreation Area, Jerusalem Forest　　耶路撒冷森林橄榄树园休闲区
- **116** Bishan- Ang Mo Kio Park and Kallang River　　新加坡加冷河——碧山宏茂桥公园
- **120** The Noues of Croix Catelan in Bois de Boulogne - Paris - France　　法国巴黎布洛涅公园一隅

124 Rike Park   Rike 公园

134 West Seoul Lake Park   首尔西部湖畔公园

140 Sydney Pirrama Waterfront Park   悉尼 Pirrama 滨海公园

150 Shangri-La Botanical Garden   香格里拉植物园

156 Hotarumibashi Park   Hotarumibashi 公园

168 Green Hills Tsuyama   津山绿丘

Cultural Park   文化公园

180 The Green Prismatic Columns / French Embassy Gardens in Tokyo   法国驻日大使馆花园

184 Sharpeville Memorial Garden   沙佩维尔纪念花园

190 The Pretoria Freedom Park   南非比勒陀利亚自由公园

196 Canadas Park   Canadas 公园

212 Würth La Rioja Museum Gardens   Würth La Rioja 博物馆花园

218 Olympic Sculpture Park   奥林匹克雕塑公园

Community Park   社区公园

228 Park of Luna   卢那公园

234 Village of Yorkville Park   约克维尔公园村

240 Submersible Garden   法国图尔斯 Submersible 花园

Central Park   CBD 公园

246 City Garden   城市花园

252 Bryant Park   布赖恩特公园

Leisure Space
Public
Natural Landscape
Sense of Experience

休闲空间
公共属性
自然景观
体验感

LEISURE AND TOURISM LANDSCAPE 休闲度假景观

TOURISM PARK 旅游度假公园 | CITY PARK 城市公园

Keywords 关键词

Costal Features 海滨特色
Architectural Details 建筑细部
Space 空间
Environment 环境

Location: Anaklia, Samegrelo-Zemo Svaneti, Georgia
Landscape Design: CMD Ingenieros
Authors: Alberto Domingo, Carlos Lázaro, Juliane Petri
Gross Floor Area: 80,400 m²

项目地点：格鲁吉亚萨梅格列罗-上斯瓦涅季亚州阿纳克利亚
景观设计：西班牙 CMD Ingenieros
设 计 师：Alberto Domingo, Carlos Lázaro, Juliane Petri
总建筑面积：80 400 m²

Anaklia Costal Park
Anaklia 海滨公园

Features 项目亮点

The design takes the regeneration of the wetland landscape resources into full consideration and sets an integration of the large-scale open space to be a modern and sustainable space with the modern technology and materials.

该公园的设计充分考虑了周边湿地的景观资源再生，将大面积的空旷区域进行整合设计，运用现代技术工艺和材料，打造了一个现代化的可持续发展空间。

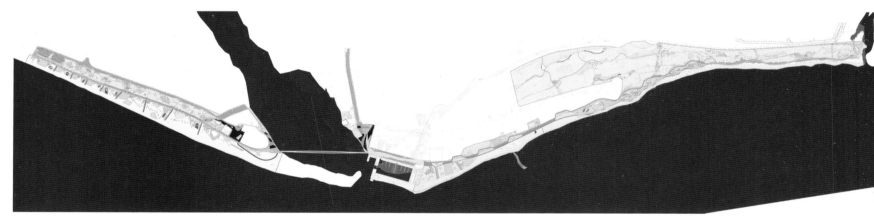

| ECOLOGY PARK 生态公园 | CULTURAL PARK 文化公园 | COMMUNITY PARK 社区公园 | CENTRAL PARK CBD公园 |

Overview

Anaklia Costal Park is a project of a new seafront park and the landscape regeneration of an extensive wetland nearby the Kolheti's Natural Park, protected by the Ramsar Convention.

The project is divided in three areas: Zone 1-Beach Resort Ganmukhuri(266,000 m²); Zone 2-Marina Anaklia(136,000 m²); Zone 3-Natural Park(401,000 m²). Zones 1 and 2 were completed in August 2011, including the integration on the master plan, the construction of the new seafront boulevard, the marina area, sport and leisure facilities, and an open-air theatre. The natural park wetland will be developed in the future until it achieves a complete surface of 803,000 m².

项目概况

该项目是一个新海滨公园,是针对受"拉尔萨穆公约"保护的Kolheti自然公园附近的广阔湿地所进行的景观再生。

项目分为三个区域:1区是Ganmukhuri海滩度假村(26.6万m²),2区是Anaklia码头(13.6万m²),3区是自然公园(40.1万m²)。1区和2区于2011年8月完成,包括总体规划的整合,新海滨大道的建设,码头区,体育和休闲设施以及露天剧场。自然公园湿地未来将发展到80.3万m²的表面积。

LEISURE AND TOURISM LANDSCAPE 休闲度假景观

TOURISM PARK 旅游度假公园　　**CITY PARK** 城市公园

Design Description

The project sets a complete definition for all the construction details for pavements, gardening, facilities, lightening, civil works, furniture, beach equipment, infrastructure design, architectural definition of the buildings for toilets, cafés, tourism information etc. There are special furniture elements and landmarks for welcome squares and meeting areas: bronze tree sculptures, bronze and ceramic pieces for the pavement, benches with the shape of a flower (identity mark of the project), and special lightening elements.

The future natural park on the south has the highest landscape quality of the project. Pavements, outdoor furniture and suggested materials for the promenade and the beach facilities have been chosen considering integration and durability at this seafront environment. The master plan reserves space for a water treatment plant which will serve hotels and the future urban area.

The Coastal Park has reached important goals on its maintenance and sustainability: effective use of local materials with the result of a low-budget-park with a high level of services and an excellent, modern image; implement of a bicycle lane for the promotion of sustainable urban transport medium; special care in the selection of materials and furniture that offers low maintenance costs and a high durability especially resistant to the maritime climate.

| ECOLOGY PARK 生态公园 | CULTURAL PARK 文化公园 | COMMUNITY PARK 社区公园 | CENTRAL PARK CBD公园 |

设计说明

项目对各个建筑细节设置了完善的定义，如路面、园艺、设施、照明、土建工程、家具、海滩设备、基础设施设计，以及厕所、咖啡厅、旅游信息等。迎宾广场和会议区还引进了特殊的家具元素和地标：青桐树雕塑，青铜和陶瓷片路面、花形板凳（该项目的特色标志）以及特殊的照明元素。

南面未来的自然公园拥有整个项目最高的景观质量。路面、户外家具和长廊材料、海滩设施的选用都考虑到了整个海滨环境的整合情况和耐久性。总体规划中保留了一定的空间作为污水处理厂的场地，将服务于酒店和未来的市区。

海滨公园在可持续发展和维护方面已取得了重大成就：有效地利用当地材料，以低预算打造了一个具有高水准服务水平和良好现代形象的公园；规划自行车道，促进可持续发展的都市交通媒介；注重材料和家具的选择，降低维修成本，提高抗海洋性气候的耐用性。

LEISURE AND TOURISM LANDSCAPE 休闲度假景观

TOURISM PARK 旅游度假公园　　CITY PARK 城市公园

| ECOLOGY PARK 生态公园 | CULTURAL PARK 文化公园 | COMMUNITY PARK 社区公园 | CENTRAL PARK CBD 公园 |

| LEISURE AND TOURISM LANDSCAPE 休闲度假景观 | TOURISM PARK 旅游度假公园 | CITY PARK 城市公园 |

Keywords 关键词

- Filigree Membrane 丝网天蓬
- Geometrical Shape 几何形状
- Membrane Construction 膜结构
- Privacy Protection 私密性

Location: Bad Wildbad, Germany
Client: Staatsbad Wildbad Bäder- und Kurbetriebs GmbH
Architects: form TL
Photography: Roland Halbe

项目地点：德国巴特维尔德巴德
客　　户：Staatsbad Wildbad Bäder- und Kurbetriebs GmbH
项目设计：form TL
摄　　影：Roland Halbe

Membrane Roof Construction Palais Thermal, Bad Wildbad, Germany

德国巴特维尔德巴德 Palais Thermal 膜结构屋顶

Features 项目亮点

The light-weight construction appears to float in the air, providing protection to privacy and against tough weather, as well as filigree membrane makes the south view which towards to the river valley clear-cut.

整个设计轻盈得如同漂浮一般，既保护客人的隐私又有效遮挡了恶劣天气，此外网状天蓬还使面临河谷和南部的风景更加清晰、有轮廓。

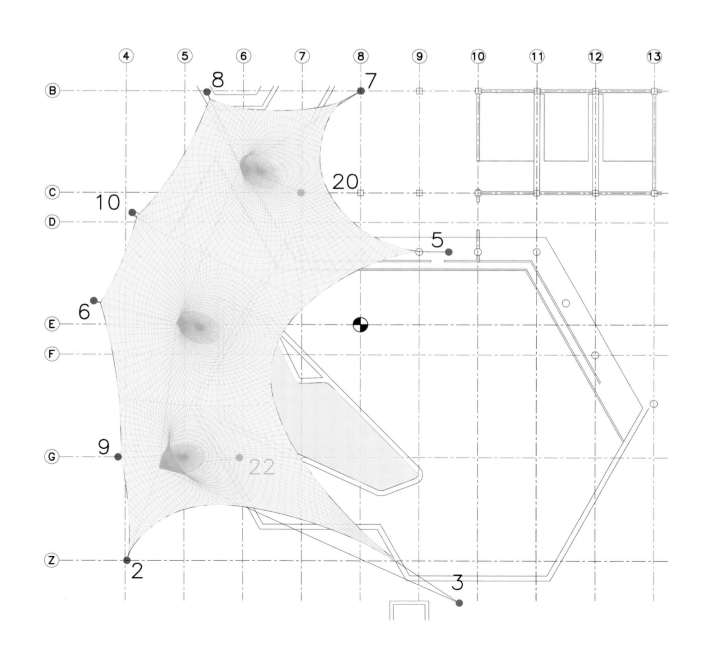

| ECOLOGY PARK 生态公园 | CULTURAL PARK 文化公园 | COMMUNITY PARK 社区公园 | CENTRAL PARK CBD公园 |

Design Description

In order to prevent people looking in from the nearby hotel and to offer sauna users privacy, the entire north-facing side of the pool is covered by a filigree membrane construction. It offers protection against the weather, as well as filigree membrane makes the south view which towards to the river valley clear-cut.

Despite the geometric and constructional challenges, or rather because of these, the light construction appears to float. Its basic structure consists of a pre-tensioned membrane with three high points and lateral anchoring. In order to offer effective protection from prying eyes, the membrane roof is tilted on its longest axis, forming some extreme geometrical shapes: the construction comprises some very steep and some very flat areas. Moreover, the organically shaped high point rings around the masts are perfectly geometrically integrated into the double-curved membrane surface, offering unhindered views up to the sky while the customers enjoying the sauna.

设计说明

为了防止周围宾馆中的旅客的窥探，同时为洗桑拿的客人提供良好的隐私保护，在泳池北面一侧设计了一个丝网的天蓬，既保护客人的隐私又有效遮挡了恶劣天气，此外天蓬还使面临河谷和南部的风景更加清晰、有轮廓。

尽管项目面临几何和建筑方面的一些挑战，但正是由于这些挑战，使整个设计轻盈得如同漂浮一般。天蓬的基本结构是一个预拉伸的膜结构，以及通过三个支撑点进行横向锚定。为了尽可能防止偷窥，天蓬向最长的轴进行倾斜，形成一个仿佛极端的几何形状。天蓬的顶部时而陡峭，时而平坦。此外，围绕桅杆进行设置的支撑点形成双曲形，使客人在享受桑拿浴的同时还可以尽情欣赏到天空的景色。

LEISURE AND TOURISM LANDSCAPE 休闲度假景观

| TOURISM PARK 旅游度假公园 | CITY PARK 城市公园 |

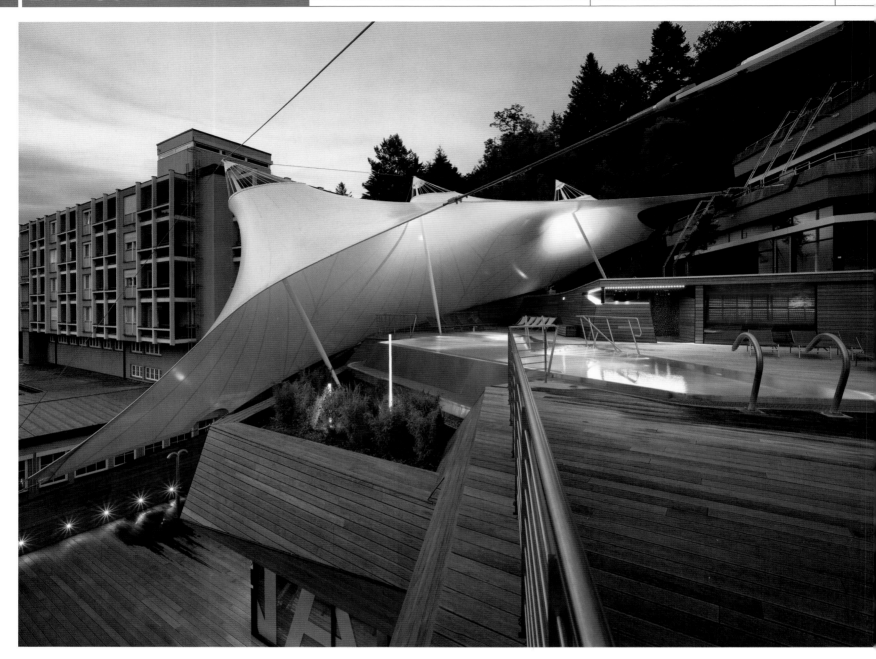

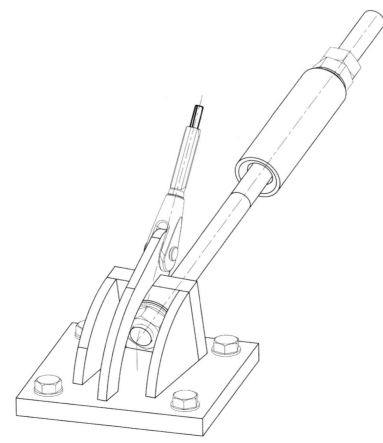

| ECOLOGY PARK 生态公园 | CULTURAL PARK 文化公园 | COMMUNITY PARK 社区公园 | CENTRAL PARK CBD公园 |

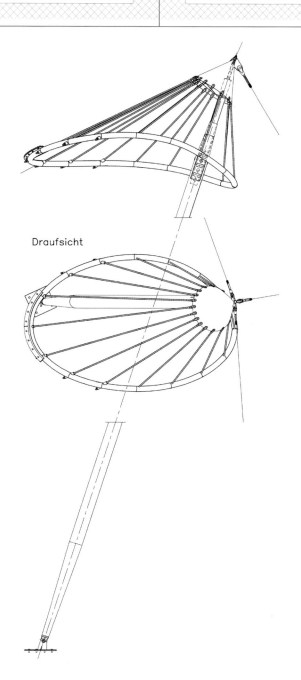

Draufsicht

| LEISURE AND TOURISM LANDSCAPE 休闲度假景观 | TOURISM PARK 旅游度假公园 | CITY PARK 城市公园 |

Keywords 关键词

Variety 功能多样

Open Space 开放空间

Harbor Features 海港特色

Innovative Technology 创新技术

Location: Los Angeles, California, USA
Client: Administration Committee of Los Angeles Port
Landscape Design: Sasaki Associates, Inc.
Area (size): 121,405.69 m²

项目地点：美国加利福尼亚州洛杉矶
客　　户：洛杉矶港口管理委员会
景观设计：Sasaki Associates, Inc.
规划面积：121 405.69 m²

Wilmington Waterfront Park
威尔明顿海滨公园

Features 项目亮点

Through integration of a series of public entertainment and activity functions which effectively separates the impact of the port, the design creates diverse public open space buffering the community and the coast.

设计通过整合一系列公共娱乐及活动功能，提供连接社区与海岸的缓冲区，有效隔开了港口的影响，打造出多元融合的公共开放空间。

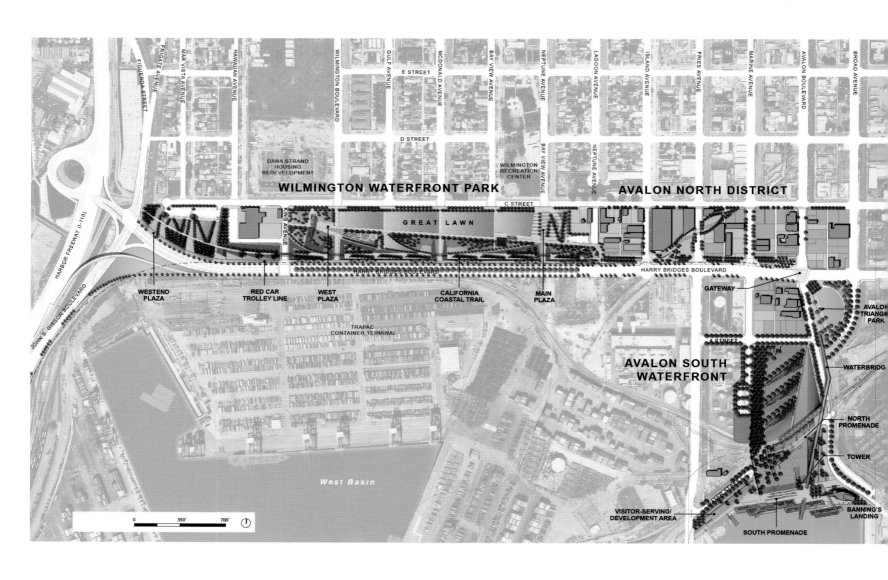

| ECOLOGY PARK 生态公园 | CULTURAL PARK 文化公园 | COMMUNITY PARK 社区公园 | CENTRAL PARK CBD公园 |

Overview

Once a part of the Pacific coastline, the Wilmington community became disconnected from the waterfront by the Port of Los Angeles—a burgeoning, diverse mix of industrial maritime facilities. After completing the Wilmington Waterfront Master Plan, Sasaki identified three open spaces for implementation: the Wilmington Waterfront Park, the Avalon North Streetscape, and the Avalon South Waterfront Park. The Wilmington Waterfront Park is the first project to be fully implemented. Built on a 30-acre(121,405.9 m²) waste industrial site, the new urban park revitalizes the community and visually reconnects it to the waterfront. The park integrates a variety of active and passive uses—informal play, public gathering, community events, picnicking, sitting, strolling, and observation—determined through an extensive community outreach process. The open space serves as a public amenity by doubling the current community open space while also buffering the Wilmington community from the extensive port operations to the south.

项目概况

威尔明顿社区曾是太平洋海岸线的一部分，但由于洛杉矶港这个迅速发展的多样性混合工业化航海设施，使其社区与滨水区分开。在完成威尔明顿滨水区总体规划后，Saskai确定了三个开放空间以供实施：威尔明顿滨水公园、阿瓦隆北部街景，以及阿瓦隆南部滨水公园。威尔明顿滨水公园是第一个完全实施的项目。新的城市公园建造在30英亩（121 405.69 m²）的工业废弃基地上，在视觉上重新与滨水区相连，并恢复了该社区。公园整合了各种积极和休憩活动功能——非正式演出、公共集会、社区活动、野炊、座椅、散步和观赏，在广泛的社区扩展过程中确定下来。开放空间作为公共娱乐设施，兼用作现有社区的开放空间，并同时为威尔明顿社区提供缓冲区，隔开南部大规模作业港口。

LEISURE AND TOURISM LANDSCAPE 休闲度假景观

TOURISM PARK 旅游度假公园

CITY PARK 城市公园

Design Description

In order to protect the community park from the port's impacts, Sasaki created a strong sculptural landform which elevates the existing planar grade of the neighborhood to 16 feet(4.88 m). This land integrates a series of multipurpose playfields with shade-dappled, gentle grass slopes. Atop the landform, the El Paseo Promenade provides a primary component of the pedestrian circuit with seating, display gardens, and a shared use pedestrian and bicycle path linked to the California Coastal Trail. Tree-lined promenades extend the park's network of pedestrian circuits and meanders, offering a variety of seating for respite, contemplation, and viewing park activities including interactive water features, an adventure playground for children, plazas for gathering and performances, and picnicking within the tree groves. Datum Walk provides a central pedestrian axis traversing the park and connecting two park pavilions. The pavilions frame outdoor rooms that offer a variety of informal seating, shade, a dry concessio public restrooms, and three flexible, formal performance venues.

Sasaki integrated sustainable design practices and innovative engineerin technologies into the overall project. Storm water management directs water primary landscape zones to promote infiltration rather than municipal treatmen demolished paving was ground and reused for paving sub-base, and all pla materials were selected as ecologically adapted, indigenous, or salt tolera and irrigated by reclaimed water. Building and site lighting highlights key pa elements, reducing energy demands and light pollution through high optic efficacy. Along the port's industrial edge, colorful planes forming the terrace wal are coated with titanium oxide [Ti02], which transforms harmful air pollutants inert organic compounds with (the use of) innovative photocatalytic technology.

| ECOLOGY PARK 生态公园 | CULTURAL PARK 文化公园 | COMMUNITY PARK 社区公园 | CENTRAL PARK CBD公园 |

设计说明

为了保护社区公园免受港口的影响，Sasaki创造了有力的雕塑感地形，将社区现有地平面升高到16英尺（4.88 m）。该基地整合了一系列多功能的游乐场地，其中有带遮阴功能的草地缓坡。在该地形顶部，El Paseo散步大道提供了主要的步行通道，其间分布着座椅、展览花园，以及一条人行和自行车共用道，与加利福利亚海岸道相连。林荫散步道扩展了公园的步行散步系统和曲径网络，沿着提供了多种多样的座椅，供休息、冥想和观看公园活动；包括互动水景、儿童探险游乐场、集会和表演广场，以及树林中的野炊活动。达顿步道提供了一条中心步行轴线，横穿公园并连接两个公园凉亭。凉亭建立了户外房间框架，提供了各种非正式座椅、遮阴、退让区、公共卫生间和3个灵活的正式表演场地。

Sasaki在整个项目中融合了可持续性设计策略和创新性工程技术。雨洪管理系统将水引入主要的景观区，促进渗透而不是流向市政污水处理；拆除的路面被回收再利用，用于铺设底基层；所有种植材料的选择都考虑了生态适应性，采用本土且耐盐性强的、可用回收水灌溉的植被。建筑和基地照明突出了关键的公园元素，通过高效光能的利用降低能源消耗和灯光污染。沿港口的工业边缘，彩色的平面形成台阶墙，涂上氧化钛，运用创新性光催化技术将有害的空气污染物转变成惰性的有机化合物。

LEISURE AND TOURISM LANDSCAPE 休闲度假景观

TOURISM PARK 旅游度假公园

CITY PARK 城市公园

| ECOLOGY PARK 生态公园 | CULTURAL PARK 文化公园 | COMMUNITY PARK 社区公园 | CENTRAL PARK CBD 公园 |

| LEISURE AND TOURISM LANDSCAPE 休闲度假景观 | TOURISM PARK 旅游度假公园 | CITY PARK 城市公园 |

Keywords 关键词

Spatial Layout 空间布局
Monumentality 纪念性
Path Network 道路网络
Landscape Design 景观设计

Location: Fortress Ehrenbreitstein, Cologne, Germany
Client: Federal Country of Rheinland-Pfalz
Landscape Design: Topotek 1
Photography: Hanns Joosten

项目地点：德国科隆 Ehrenbreitstein 堡垒
客　　户：普法尔茨州联邦
景观设计：Topotek 1 设计工作室
摄　　影：Hanns Joosten

Fortress Ehrenbreitstein
Ehrenbreitstein 堡垒

Features 项目亮点

The spatial layout is creatively transferred while keeping the unique monumentality and organizing the whole surface through newly built path network.

创造性地转变空间布局，保持了项目独特的纪念性质，并通过新建的道路网络将整个平面有序地组合在一起。

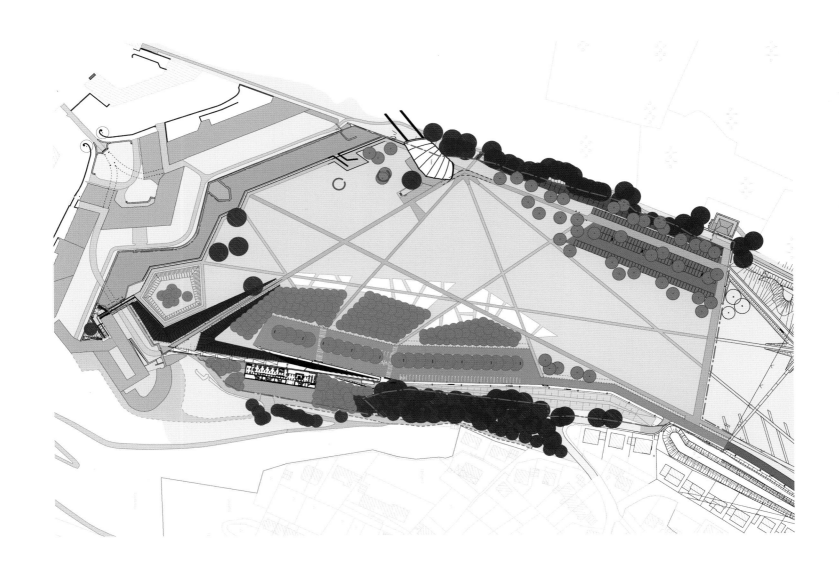

| ECOLOGY PARK 生态公园 | CULTURAL PARK 文化公园 | COMMUNITY PARK 社区公园 | CENTRAL PARK CBD公园 |

Overview

The huge stronghold, overlooking the confluence of Rhine and Mosel rivers is an important national monument. As a tourist facility of the reconstruction project of the north, the historical plateau is developed as a museum park and the spatial qualities of the site brought into context with the wider surrounding view.

项目概况

这座巨大的要塞是一个重要的国家纪念碑，俯瞰莱茵河和摩泽尔河的交汇处。作为北部重建工程中重要的游客设施，这座历史上的高地被建成了一个博物馆公园，空间特征使得此处可以尽享周围宽阔壮丽的风光。

LEISURE AND TOURISM LANDSCAPE 休闲度假景观

TOURISM PARK 旅游度假公园

CITY PARK 城市公园

Design Description

The general design concept reorganizes the historic fortress plateau in a spatial and dramaturgic way into the new main entrance of the Fortress Ehrenbreitstein. The car entrance and the parking lot are placed at the edge of the plateau. Through this the big space keeps its monumentality and stages the silhouette of the Fortress to the north. On top of the former baroque design construction is a wide grass field laid out which makes out the background for the experience of the Fortress ensemble. A new network of path axes orders the whole plane and is connected with the existing system of pathways.

| ECOLOGY PARK 生态公园 | CULTURAL PARK 文化公园 | COMMUNITY PARK 社区公园 | CENTRAL PARK CBD公园 |

设计说明

基本的设计理念将这座历史堡垒在空间上以戏剧化的方式转变成Ehrenbreitstein堡垒的新主入口。车辆入口和停车场位于高地的边缘。通过这样的安排方式使整个大空间依然保持着其独特的纪念性，同时堡垒的轮廓被上升到了北方。以前的巴洛克式设计方案是一片巨大的草坪，作为整个堡垒区域的大背景。而新的道路网络不仅将整个平面有序地组合在一起，同时还可以与旧的通道连接起来。

LEISURE AND TOURISM LANDSCAPE 休闲度假景观

TOURISM PARK 旅游度假公园

CITY PARK 城市公园

| ECOLOGY PARK 生态公园 | CULTURAL PARK 文化公园 | COMMUNITY PARK 社区公园 | CENTRAL PARK CBD 公园 |

LEISURE AND TOURISM LANDSCAPE 休闲度假景观

TOURISM PARK 旅游度假公园

CITY PARK 城市公园

| ECOLOGY PARK 生态公园 | CULTURAL PARK 文化公园 | COMMUNITY PARK 社区公园 | CENTRAL PARK CBD公园 |

| LEISURE AND TOURISM LANDSCAPE 休闲度假景观 | TOURISM PARK 旅游度假公园 | CITY PARK 城市公园 |

Keywords 关键词

Imagination 想象力
Spatial Experience 空间体验
Alien Temperament 异域风情
Cultural Communication 文化交流

Location: Xi'an, Shanxi, China
Landscape Design: Topotek 1
Photography: Wang Geng, TOPOTEK 1, Wang Xiangrong

项目地点：中国陕西西安
景观设计：德国 Topotek 1 事务所
摄　　影：Wang Geng, TOPOTEK 1, Wang Xiangrong

The Big Dig, Xi'an Garden Show

西安公园展——大地洞

Features 项目亮点

The Big Dig is taken as a loud speaker to activate connections with the garden and bring the sound from the other end of the world to visitors, which enhances the experience of landscape.

大地洞被当作一个扩音器，以激发其与花园之间的联系，并为游览者带来世界另一端的声音，极大地增强了景观的体验性。

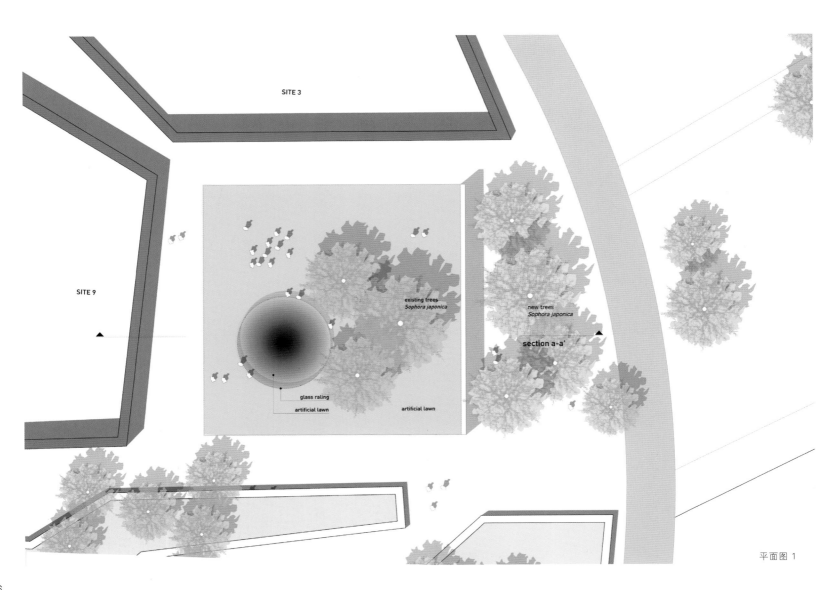

平面图 1

| ECOLOGY PARK 生态公园 | CULTURAL PARK 文化公园 | COMMUNITY PARK 社区公园 | CENTRAL PARK CBD公园 |

Overview

There is an international dream fantasizing about the other side of the world. Perhaps it results from being the furthest possible distance? A completely foreign culture? An unknown geography? A common childhood warning of, "If you keep digging, you'll dig all the way to China" is often an introductory presentation of this dream. The whole idea is a spirit of adventure and whimsy, and the unending curiosity of what can be found on the other side.

In the garden, architects have created the result of this overzealous dig: the point that the hole emerges in China. At this hole they capture a precise point where one stands on the edge of one world and another, wondering what is possible of the other side? As tradition, a garden is a place that transfers someone into a 'foreign' space: from inside to outside, from city to nature, from one culture to another. This garden is the cusp at which two worlds are colliding, a foreign world entering China, defined by the visitor's imagination.

项目概况

人心中都有一个世界梦，幻想着世界另一端的景象。也许是因为距离产生美，也许是对异域文化的向往，或是对未知地理的着迷……孩童时代，大人们会告诉我们，如果我们一直挖地洞，就会挖到中国去。这个"世界梦"实际上反映了人类的冒险探索精神和对地球另一端永无止境的好奇心。

本花园中，设计者创造了挖地洞的结局：地洞到达中国的那一点。在这个地洞中，捕捉到精确的一个位置，这是世界两端交界的地方，游客在此可想象世界另一端的景象。从传统意义上来说，花园是将人们带入一个"外来"空间的地方：从室内到室外，从城市到自然，从一种文化到另一种文化。本花园是世界两端交界的尖端点，游客可尽情发挥他们的想象力，幻想中国以外的异域风情。

LEISURE AND TOURISM LANDSCAPE 休闲度假景观

TOURISM PARK 旅游度假公园

CITY PARK 城市公园

| ECOLOGY PARK 生态公园 | CULTURAL PARK 文化公园 | COMMUNITY PARK 社区公园 | CENTRAL PARK CBD公园 |

平面图 2

LEISURE AND TOURISM LANDSCAPE 休闲度假景观

TOURISM PARK 旅游度假公园

CITY PARK 城市公园

Design Description

Architects also see this hole as not just an object, but a spatial exercise to consider the superficial surface. Topotek1 has a tradition of provoking the very idea of a surface, and what it could mean in different contexts. The garden site provided to us is immediately thought of as a flat surface. The sunken hole is a physical action to emotionally bring the visitor to the global context. As the hole is to have just 'emerged', the site is visible as an artificial grass carpet which is sucked down into the hole. Existing trees have remained, keeping the site as seemingly untouched. Underneath the grass, the 10 m wide hole is a concrete shell structure excavated into the earth. A glass railing surrounds the hole… one doesn't want to get too curious…

With the garden, architects take the literal shape of the hole and see it as megaphone. A megaphone is used to address large outdoor audiences or to spe to people from a long distance. In our case they needed such an instrument emit audio captured from the other side of the world. Selecting Argentina, t United States, Sweden, and Germany as the foreign locations, one will be able hear the other side of the world as they near the hole. There will be soundtrac of the life on the other side: cows from the pampas of Argentina, commute rushing among transit through New York City, the maritime life of Stockholm, a layers of history so audible among the streets of Berlin. These soundtracks piq the imagination of the visitors, transferring them away from China, away from t garden, away from the hole, and to the other side of the world.

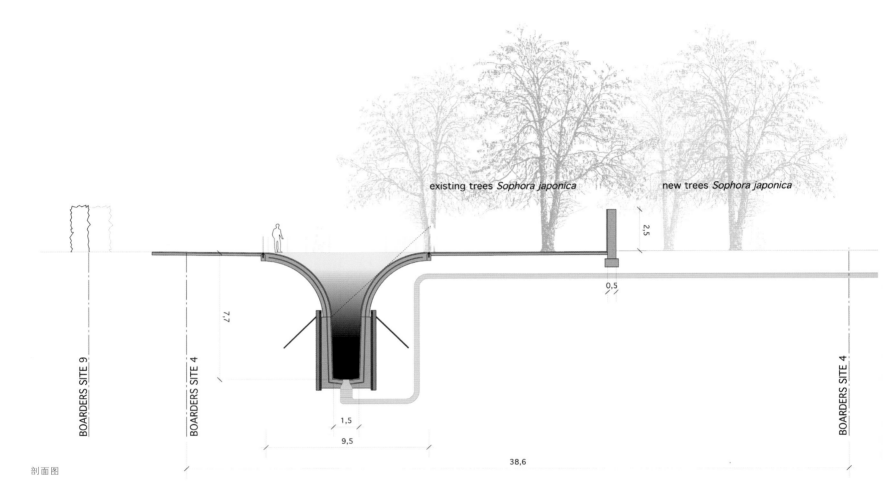

剖面图

| ECOLOGY PARK 生态公园 | CULTURAL PARK 文化公园 | COMMUNITY PARK 社区公园 | CENTRAL PARK CBD公园 |

设计说明

个地洞不仅是一个物体，同时也为游客提供仔细观察表层地面的空间体验。Topotek1 有研究地表层在不同环境中的意义的传统。花园所在场所地势平坦，下沉的地洞借助物理行为，激发游客感，为游客营造跨国环境氛围。地洞若隐若现，整个场所宛若一张人工大草皮，被吸入地洞之中。计者保留了原有的树木，以维持场所的原状。地洞则是草皮下方的一个 10 m 宽混凝土壳层结构，周设有玻璃扶手，吸引游客前来观看。

为进一步激发与花园相关的情感，设计师还利用地洞的外观，把它当作一个扩音器。扩音器通常是在户外现场演说或者远距离对话时使用。这座景观需要这样一个设备，用来播放来自世界另一端的音频文件。设计者选择了阿根廷、美国、瑞典和德国作为外国站点，当游客接近地洞时，能够聆听到世界另一端的声音：那是来自阿根廷彭巴斯草原上的奶牛，这是来自纽约交通干道上的通勤者，还有瑞典斯德哥尔摩的海员，以及历史悠久的德国街区。这些声音能够激发游客的想象力，把他们带出中国，离开花园，离开地洞，到达世界的另一端。

LEISURE AND TOURISM LANDSCAPE 休闲度假景观

TOURISM PARK 旅游度假公园

CITY PARK 城市公园

Keywords 关键词

| Altitude 海拔 |
| Material 材料 |
| Landscape Element 景观元素 |
| Natural Environment 自然环境 |

Location: Romsdalen - Geiranger Fjord, Norway
Client: The Norwegian Public Roads Administration
Architects: Reiulf Ramstad Architects (RRA)

项目地点：挪威盖朗尼尔峡湾 - 罗姆斯达伦
客　　户：挪威公共道路建设局
景观设计：挪威 Reiulf Ramstad Architects 建筑事务所

Trollstigen Tourist Route Project

特洛斯蒂国家观光旅游线路

Features 项目亮点

The project designs a stretching path up to the mountain, enhances to adap the Trollstigen plateau's location and experience of the nature, which make the vocation space more unique.

设计一条绵延的上山通道，加深了游客对该地高原位置的适应及对该地自然的体验，让这一度假空间更具独特性。

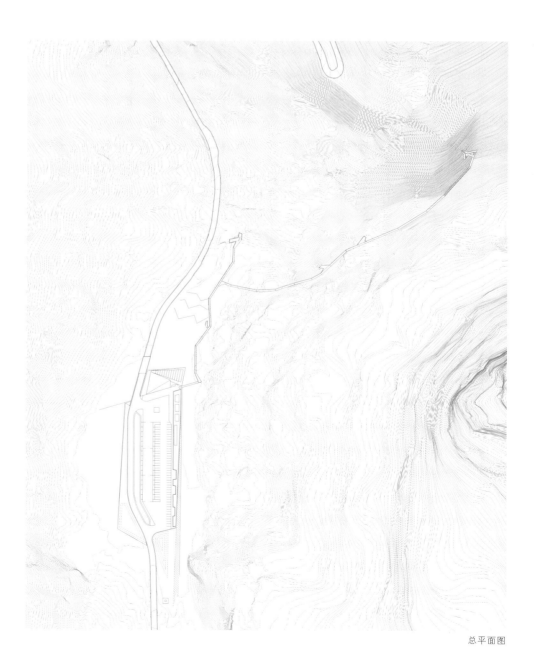

总平面图

| ECOLOGY PARK 生态公园 | CULTURAL PARK 文化公园 | COMMUNITY PARK 社区公园 | CENTRAL PARK CBD 公园 |

Overview

Located on Norway's west coast, Trollstigen is perched within a dramatic pass between the deep fjords that characterize the region. This panoramic site can only be visited and constructed in summer, due to steep slopes and severe weather.

项目概况

特洛斯蒂位于挪威的西海岸,从这个地区极富特色的深海湾之间巧妙地穿过。这个风景区景色秀美,但山势陡峭不易到达,并且气候恶劣,只有在夏季可以接待游客,也只有在夏季可以进行施工建设。

LEISURE AND TOURISM LANDSCAPE 休闲度假景观

TOURISM PARK 旅游度假公园　　　　**CITY PARK** 城市公园

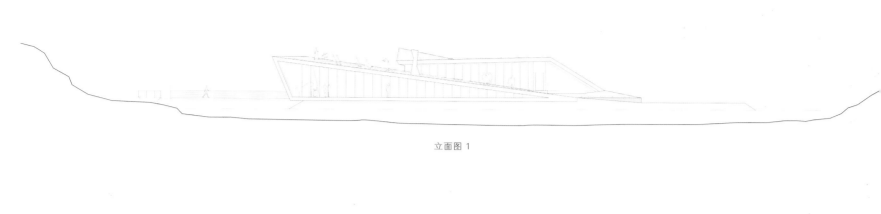

立面图 1

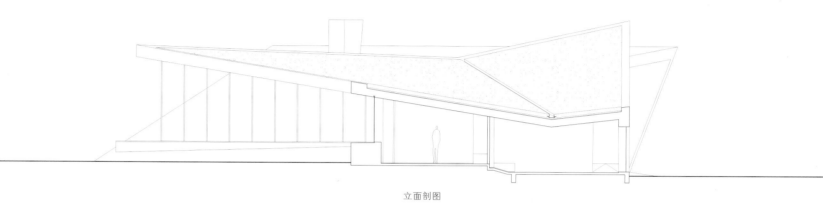

立面剖图

| ECOLOGY PARK 生态公园 | CULTURAL PARK 文化公园 | COMMUNITY PARK 社区公园 | CENTRAL PARK CBD 公园 |

| LEISURE AND TOURISM LANDSCAPE 休闲度假景观 | TOURISM PARK 旅游度假公园 | CITY PARK 城市公园 |

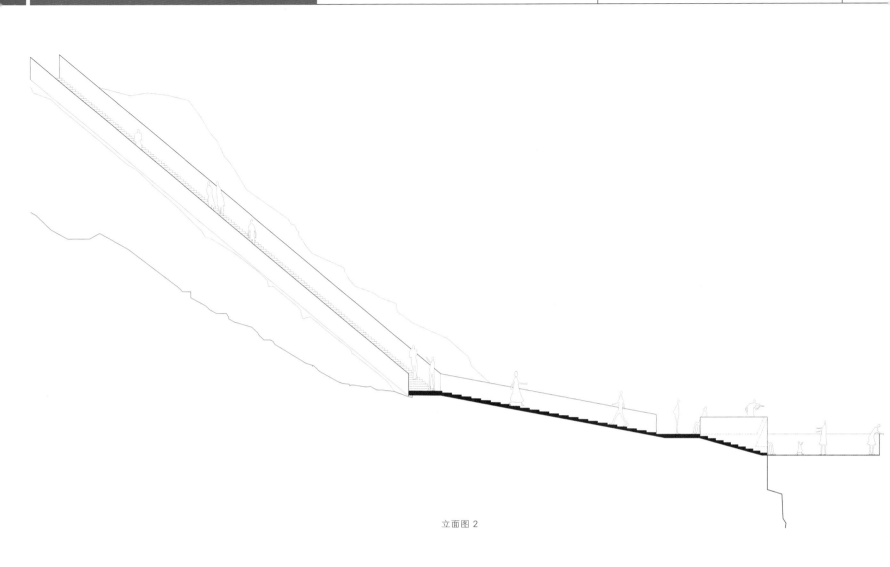

立面图 2

| ECOLOGY PARK 生态公园 | CULTURAL PARK 文化公园 | COMMUNITY PARK 社区公园 | CENTRAL PARK CBD公园 |

Design Description

Despite—or perhaps because of—the inaccessible nature of the site, the project entails designing an entire visitor environment ranging from a mountain lodge with restaurant and gallery to flood barriers, water cascades, bridges, and paths to outdoor furniture and pavilions and platforms meant for viewing the scenery. All of these elements are molded into the landscape so that the visitor's experience of place seems even more intimate. The architectural intervention is respectfully delicate, and was conceived as a thin thread that guides visitors from one stunning overlook to another.

The RRA project will enhance the experience of the Trollstigen plateau's location and nature. Thoughtfulness regarding features and materials will underscore the site's temper and character, and well-adapted, functional facilities will augment the visitor's experience. The architecture is to be characterised by clear and precise transitions between planned zones and the natural landscape. Through the notion of water as a dynamic element - from snow, to running and then falling water, and rock as a static element, the project creates a series of prepositional relations that describe and magnify the unique spatiality of the site.

设计说明

正是因为这个地方难以接近，项目需在这里铺设一条绵延的上山通道，海拔在游人的行走中慢慢抬升，不会有突增的不适感；另外，在峡湾之上还建造了一个伸展在半空中的观景台，游人可以全方位地欣赏到当地的美景。项目的整体设计范围包括带餐厅的山上小屋、通往防洪坝的走廊、瀑布、小桥、连接室外的小道、亭子，还有观景平台。所有这些元素共同塑造成的景观，使游客们能与景观更加亲近。建筑的引入十分微妙，犹如一条纤细的丝线，牵引着游客从一个美景跳到下一个美景。

设计师的设计加深了旅客对特洛斯蒂高原的位置与自然的体验，对特征和材料的考虑强调了这个地区的特点和秉性，其适应性强的功能设施能够让游客更加舒适。建筑师希望能让人工区域与自然景观之间的过渡更加清晰明确。通过各种形态的水（雪、流水、瀑布）和石头等动态、静态元素的运用，他们找到了一系列能描述和放大这个地区的独特空间特性的对应关系。

节点图1

LEISURE AND TOURISM LANDSCAPE 休闲度假景观

TOURISM PARK 旅游度假公园

CITY PARK 城市公园

| ECOLOGY PARK 生态公园 | CULTURAL PARK 文化公园 | COMMUNITY PARK 社区公园 | CENTRAL PARK CBD 公园 |

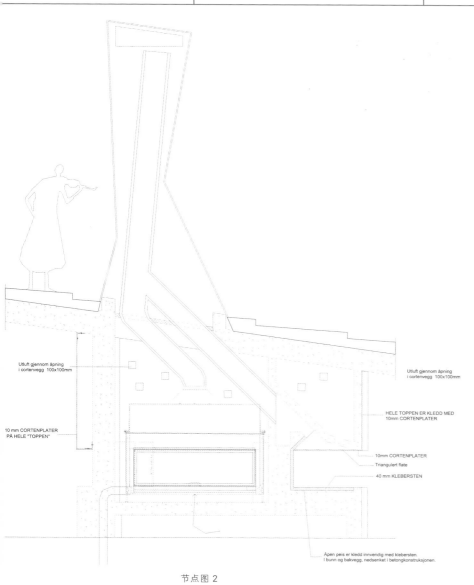

节点图 2

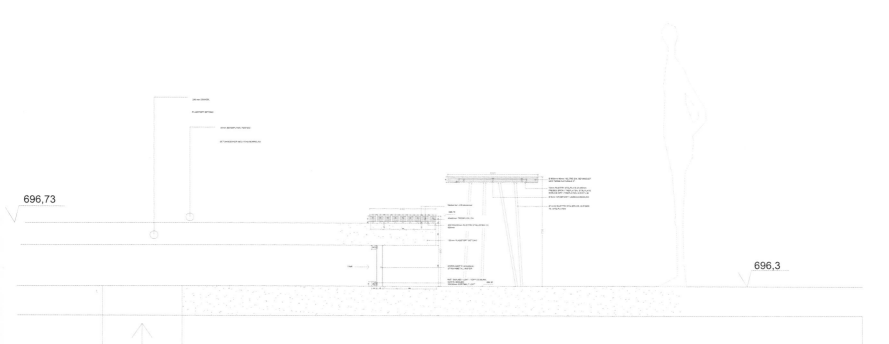

节点图 3

LEISURE AND TOURISM LANDSCAPE 休闲度假景观

TOURISM PARK 旅游度假公园

CITY PARK 城市公园

| ECOLOGY PARK 生态公园 | CULTURAL PARK 文化公园 | COMMUNITY PARK 社区公园 | CENTRAL PARK CBD 公园 |

LEISURE AND TOURISM LANDSCAPE 休闲度假景观

TOURISM PARK 旅游度假公园

CITY PARK 城市公园

| ECOLOGY PARK 生态公园 | CULTURAL PARK 文化公园 | COMMUNITY PARK 社区公园 | CENTRAL PARK CBD 公园 |

Urban
Relaxation Site
Sharing
Functionality

城市属性
休憩场所
共享性
功能性

LEISURE AND TOURISM LANDSCAPE 休闲度假景观

TOURISM PARK 旅游度假公园 CITY PARK 城市公园

Keywords 关键词
Play Space 游戏空间
Public Green Space 公共绿地
Roof Garden 屋顶花园
Abstracted Art 抽象艺术

Location: Darling Harbour, NSW, Sydney, Australia
Client: Lend Lease and Sydney Harbour Foreshore Authority
Design Team: ASPECT Studios
Area: 15,000 m²
Photography: Florian Groehn, John Marmaras, Hamish Ta-me

项目地点：澳大利亚新南威尔士州悉尼达令港
客　　户：Lend Lease 集团以及悉尼海港委员会
设计团队：ASPECT Studios 澳派景观设计工作室
面　　积：15 000 m²
摄　　影：Florian Groehn, John Marmaras, Hamish Ta-me

Darling Harbour
达令广场

Features 项目亮点

With "water" as the theme, it creates a unique feature, innovative and highly interactive play space in abstracted art form.

本案以水为主题，通过抽象艺术的形式，营造了一个独具特色、富含挑战、互动创新的游戏空间。

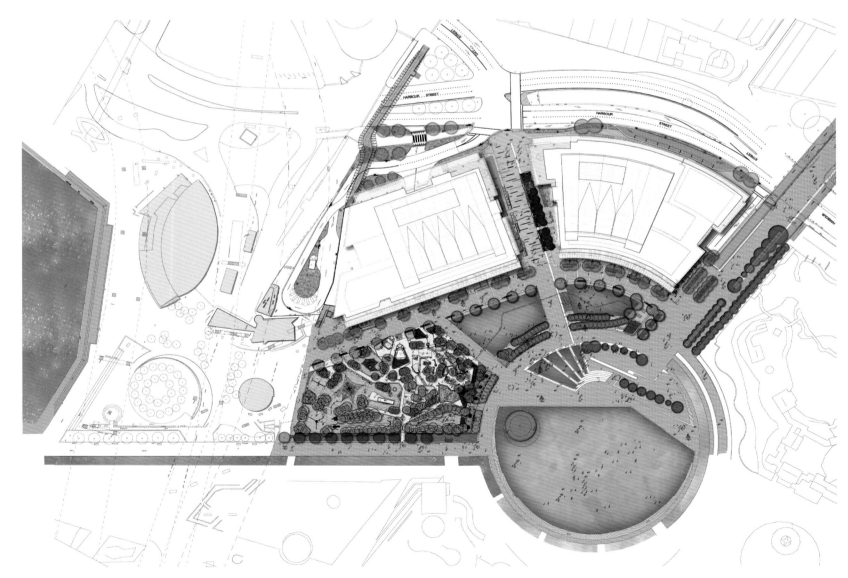

Master Plan

| ECOLOGY PARK 生态公园 | CULTURAL PARK 文化公园 | COMMUNITY PARK 社区公园 | CENTRAL PARK CBD 公园 |

Overview

The 15,000 m² project includes a retail terrace, public park, two 6-star commercial buildings, a new pedestrian street, and its centrepiece - at 4,000 m² is the largest children's play space in the Sydney CBD. The unique water play area brings Darling Harbour's industrial heritage to the fore and establishes Darling Harbour as a regional destination.

项目概况

项目占地 15 000 m²，包括一片零售商业区、一个公园、两座六星级的商业楼宇、一条新建步行街。项目中心是一座 4 000 m² 的儿童游乐场，也是悉尼 CBD 内最大的儿童游乐场。为了突出达令港工业港口的特色，游乐场以"水"为主题，打造一系列具有创意的互动游戏空间。此项目的成功无疑将会为达令港湾这一澳大利亚最吸引人的旅游胜地注入新的活力。

LEISURE AND TOURISM LANDSCAPE 休闲度假景观

TOURISM PARK 旅游度假公园   **CITY PARK** 城市公园

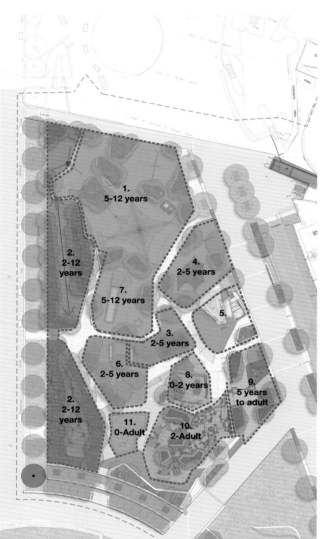

KEY

1. **Senior dynamic play area (5-12 years)**
   - Large net climbing structure
   - 4 x large swings
   - 1 x rotating swing
   - Twisted rope bridge
   - Livistona australis
   - Howea fosteriana

2. **Forest exploration area (2-12 years)**
   - Flying fox
   - Horizontal rope structure
   - Balancing beams
   - Salvaged sandstone blocks
   - Deck
   - Corymbia maculata

3. **Mogul terrain zone (2-5 years)**
   - Mounded rubber 'soft fall' terrain
   - Livistona australis
   - Howea fosteriana

4. **Junior dynamic play area (2-5 years)**
   - 2 x junior swings
   - Medium sized junior net climbing structure
   - Livistona australis
   - Howea fosteriana

5. **Observation Zone**
   - Kiosk
   - Toilets (1 x accessible with baby change, 2 x unisex)
   - Bubblers
   - Shade structure

6. **Junior slide and sand play area (2-5 years)**
   - 2 x slides
   - 2 x sand diggers
   - Livistona australis
   - Howea fosteriana

7. **Senior slide and spinning area (5-12 years)**
   - 1 x wide custom slide on rubber mound
   - 1 x spinning element
   - 1 x climbing wall
   - Tunnels through mound
   - Livistona australis
   - Howea fosteriana

8. **Baby toddler and play area (0-2 years)**
   - Sand pit
   - Carved sandstone with 'puddles'
   - Shade structure

9. **Pumping station (5 years-adult)**
   - 7 x hand pumps
   - 1 x water wheel
   - Channels in ground surface
   - Troughs above ground
   - Livistona australis
   - Howea fosteriana
   - Corymbia maculata

10. **Water channels area (2-Adult)**
    - Water channels with sluice gates and flow diverters
    - Interactive water moving elements
    - Water scooping wheel
    - Livistona australis
    - Howea fosteriana

11. **Synchronised jets area (0-Adult)**
    - Series of synchronised jets
    - Decorative applied finishes to concrete ground surface

Design Description

Two bold pedestrian north–south and east–west link connect the precinct to Sydney city, Chinatown and Cockl Bay. Activated by cafes and retail activities, they become th framework for the public domain design. The second key t transforming the Darling Harbour into a high quality family centric precinct has been to recognise that its public space must support family activities and allow social interactio between families and other users.

Darling Harbour's unique play space creates an adventurous innovative and highly interactive play experience. Its desig recalls Darling Harbour's historic waterfront landscap setting by creating an intricately detailed abstracted rive environment. Extensive research led to beautifully crafte water play elements being sourced from Germany an used in Australia for the first time. The water play area i complimented by a 'dry' playground space that features san pits, a flying fox, giant climbing nets and huge family slide built for groups to use together.

ASPECT Studios also designed two green roof terraces an community gardens as part of the precinct's new 6 star rate Commonwealth Bank Australia building and for use by th 6,000 bank staff. The roof top comprises an informal indoor-outdoor meeting area, and a breakout and entertainmen facility, and staff can participate in community gardenin activities, barbecue, eat, work or relax in the garden setting on the western edge of the buildings.

| ECOLOGY PARK 生态公园 | CULTURAL PARK 文化公园 | COMMUNITY PARK 社区公园 | CENTRAL PARK CBD公园 |

设计说明

达令广场成功地打造了南北向、东西向两条人行走廊，连接起市中心、唐人街和海扇湾。城市居民与游客可以沿街购物、观光、品尝美食，尽情享受休闲生活。这两条人行走廊也组成了公共空间规划设计的主要框架。达令港改造的另一关键点是让城市公共绿地空间更好地为家庭活动服务，促进城市居民间的交流互动，进而使整个社会更为和谐，更有凝聚力。

达令广场是一个独具特色、富含挑战、互动创新的游戏空间。设计从达令港的工业历史出发，通过抽象的艺术形式，在游乐场再现澳大利亚各大河流的景象。本项目广泛使用了从德国引进的制作精美的儿童水上游乐器材，这些器材是在澳大利亚境内首次使用。此外，水乐园中还有一个"干"的游乐园，由大沙池、滑道、攀爬网和大型的滑梯等组成，可以供多个家庭尽情游戏。

此外，澳派还为达令港两栋新的六星级建筑（澳大利亚联邦银行）打造了两个绿色屋顶花园以及企业文化花园，为联邦银行6 000名职员创建了办公休闲的好去处。屋顶花园包括一个室内外相融合的聚会区域、一个小型休闲聚会空间，以及丰富的休闲娱乐设施。同时，联邦银行的职员可以到企业文化花园种植、摆弄花草、烧烤、用餐、休闲放松。

LEISURE AND TOURISM LANDSCAPE 休闲度假景观

| TOURISM PARK 旅游度假公园 | CITY PARK 城市公园 |

| ECOLOGY PARK 生态公园 | CULTURAL PARK 文化公园 | COMMUNITY PARK 社区公园 | CENTRAL PARK CBD公园 |

LEISURE AND TOURISM LANDSCAPE 休闲度假景观

| TOURISM PARK 旅游度假公园 | CITY PARK 城市公园 |

| ECOLOGY PARK 生态公园 | CULTURAL PARK 文化公园 | COMMUNITY PARK 社区公园 | CENTRAL PARK CBD 公园 |

LEISURE AND TOURISM LANDSCAPE 休闲度假景观

TOURISM PARK 旅游度假公园

CITY PARK 城市公园

| ECOLOGY PARK 生态公园 | CULTURAL PARK 文化公园 | COMMUNITY PARK 社区公园 | CENTRAL PARK CBD公园 |

| LEISURE AND TOURISM LANDSCAPE 休闲度假景观 | TOURISM PARK 旅游度假公园 | CITY PARK 城市公园 |

Keywords 关键词

| Resting Space 休息空间 |
| Tower 塔楼 |
| Landscape 景观视线 |
| Material 材料 |

Solberg Rest Station

索伯格休息站

Location: Sarpsborg, Stfold, Norway
Architects: Saunders Architecture
Area: 2,000 m²
Photography: Bent Rene Synnevaag

项目地点：挪威东福尔郡萨尔普斯堡
景观设计：挪威 Saunders Architecture 建筑事务所
面　　积：2 000 m²
摄　　影：Bent Rene Synnevaag

Features 项目亮点

The starting point of the project is around the resting space that based on the original architecture style, it builds tower buildings, gardens and pavilions to give the whole rest station good environment and open field of vision.

设计的出发点均是围绕着休息空间展开，在原有建筑环境的风格下，塑造了诸如塔楼、庭院、亭子等景观小品，让整个休息站环境良好、视野开阔。

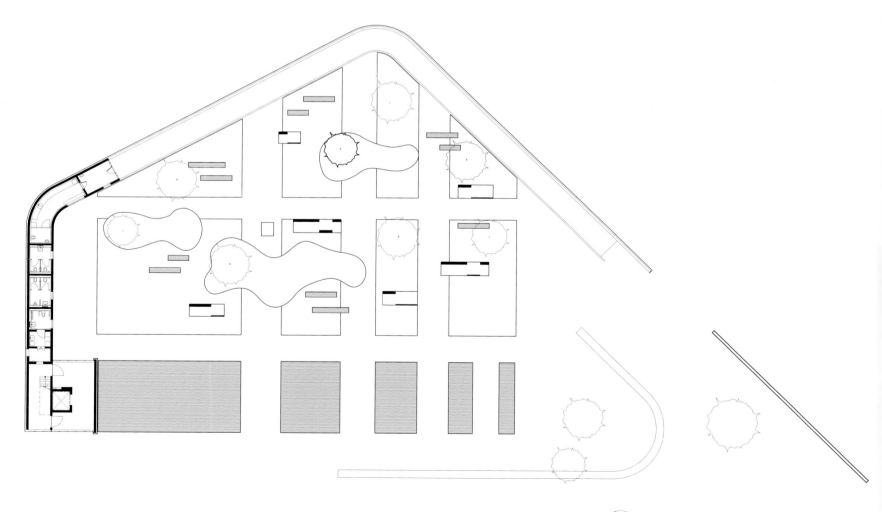

总平面图

| ECOLOGY PARK 生态公园 | CULTURAL PARK 文化公园 | COMMUNITY PARK 社区公园 | CENTRAL PARK CBD公园 |

Overview

As Sarpsborg is one of the first tastes of Norway the travelers from Sweden experience, it was important for the client that they would be able to slow down and spend time discovering the surrounding nature. The local forest and coastline form a beautiful, yet largely unknown part of the country. The neighboring highway's speed and noise only enhance the traveler's need for a break and re-connection with nature, so a green resting space was on the top of the list. A low walled ramp spirals around the rest area, defining the 2,000 m² area's limits, while spring-flowering fruit trees adorn the courtyard. Saunders and graphic designer Camilla Holcroft cooperated to design seven small pavilions in the yard, "Bronze Age" exhibition info show on the local petroglyph, it also extended to the walls of the ramp.

项目概况

萨尔普斯堡是瑞典游客进入挪威最先路过的地方，这一点对客户很重要，因为这让游客能放慢脚步，花时间打量周围的自然环境。当地的森林和海岸线成了挪威的一处旅游胜地，不过这里大多数地区还不为人知。附近高速公路的车速和噪音更加让旅行者想休息一下，重新投入大自然的怀抱，因此设计的首要元素就是一个绿色的休息空间。两侧带低矮围墙的螺旋坡道环绕着休息区，圈出了 2 000 m² 的场地范围，而春季开花的果树点缀着这个庭院。Saunders 与平面设计师 Camilla Holcroft 合作在院子里设计了七个小亭子，在当地的岩石雕刻上展示名为"铜器时代"的展览信息，还延续到了坡道的围墙上。

LEISURE AND TOURISM LANDSCAPE 休闲度假景观

TOURISM PARK 旅游度假公园

CITY PARK 城市公园

Design Description

The flatness of the landscape meant that the beauty of the surrounding nature could only be enjoyed from a certain height, so the creation of a tower quickly became a main part of the brief. The ramp's asymmetrical walls rise from 0 – 4 m, then forms a 30 m simple nine-storey-tall structure on the site's northern edge, including only a staircase and an elevator. Named Solberg (which translates into 'sun mountain'), the tower's aerial views towards the nearby coastline, the Oslo fjord are truly dramatic and impressive.

Finally, the design's style and aesthetic were developed in relation to the environs' existing architecture; minimal and geometrical contemporary shapes were chosen, contrasting the local farming villages' more traditional forms. The main materials used were beautifully-ageing CorTen steel for the exterior walls and warm oiled hard wood for the courtyard's design elements and information points. Local slate and fine gravel pave the ground level.

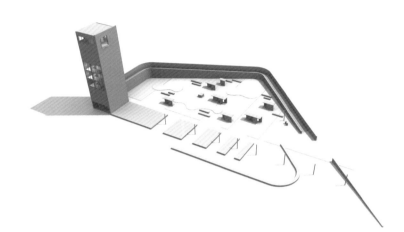
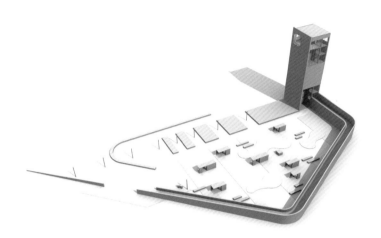

| ECOLOGY PARK 生态公园 | CULTURAL PARK 文化公园 | COMMUNITY PARK 社区公园 | CENTRAL PARK CBD 公园 |

设计说明

览无遗的美景意味着周围的大自然之美只有在一定的高度才能欣赏得到，因此在项目中设计一座楼很快成为重点。斜坡的不对称围墙从地面上升到4 m高，进而在场地北侧形成了一个简单的九高结构，高30 m，其中仅有一部楼梯和电梯。塔楼被命名为"索伯格"，意为"太阳山"，从塔上可俯瞰附近的海岸线以及奥斯陆峡湾，风景如画，令人难忘。

最后，就环境原有的建筑风格确定了本次设计的风格和审美取向，选用了极为简洁的当代几何造型，与当地农庄较为传统的建筑风格形成了对比。主要材料有用于外墙、随着时光老化得充满美感的耐候钢和用于庭院设计元素与信息板的涂有暖色颜料的硬木，当地的石板和细砾石用来铺地。

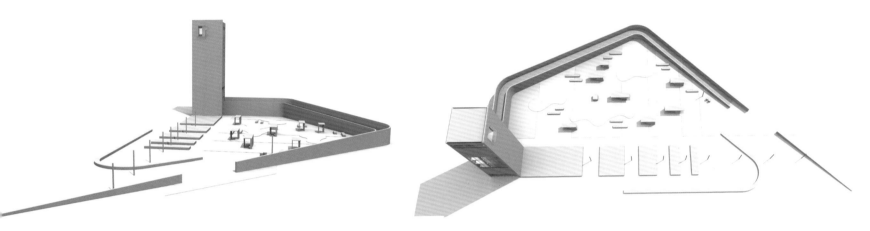

LEISURE AND TOURISM LANDSCAPE 休闲度假景观

TOURISM PARK 旅游度假公园

CITY PARK 城市公园

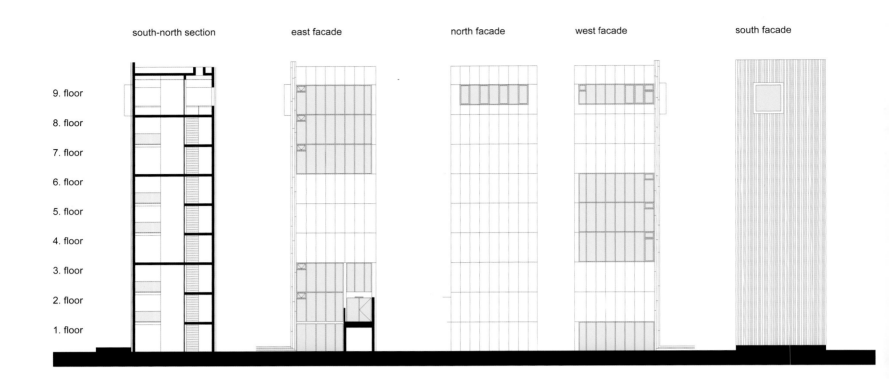

south-north section | east facade | north facade | west facade | south facade

9. floor
8. floor
7. floor
6. floor
5. floor
4. floor
3. floor
2. floor
1. floor

1. floor plan

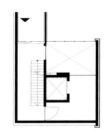
2. floor plan

3. floor plan

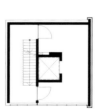
4. floor plan

5. floor plan

6. floor plan

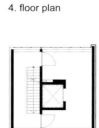
7. floor plan

8. floor plan

9. floor plan

| ECOLOGY PARK 生态公园 | CULTURAL PARK 文化公园 | COMMUNITY PARK 社区公园 | CENTRAL PARK CBD 公园 |

LEISURE AND TOURISM LANDSCAPE 休闲度假景观

| TOURISM PARK 旅游度假公园 | CITY PARK 城市公园 |

| ECOLOGY PARK 生态公园 | CULTURAL PARK 文化公园 | COMMUNITY PARK 社区公园 | CENTRAL PARK CBD公园 |

LEISURE AND TOURISM LANDSCAPE 休闲度假景观

TOURISM PARK 旅游度假公园

CITY PARK 城市公园

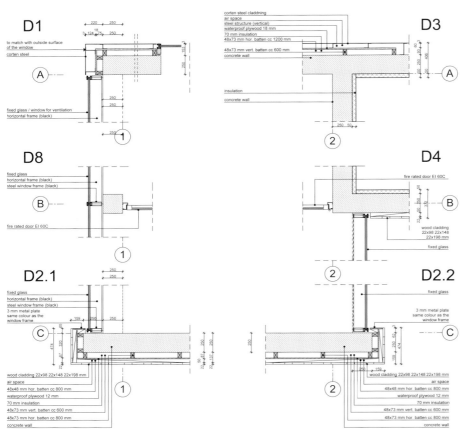

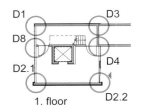

| ECOLOGY PARK 生态公园 | CULTURAL PARK 文化公园 | COMMUNITY PARK 社区公园 | CENTRAL PARK CBD 公园 |

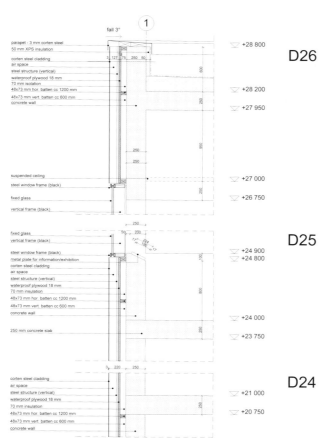

D26

D25

D24

east-west section
from south

LEISURE AND TOURISM LANDSCAPE 休闲度假景观

TOURISM PARK 旅游度假公园

CITY PARK 城市公园

| ECOLOGY PARK 生态公园 | CULTURAL PARK 文化公园 | COMMUNITY PARK 社区公园 | CENTRAL PARK CBD公园 |

| LEISURE AND TOURISM LANDSCAPE 休闲度假景观 | TOURISM PARK 旅游度假公园 | CITY PARK 城市公园 |

Keywords 关键词

Landscape Preview 景观预览
Panoramic View 全景视角
Model Area 样板区
Space Level 空间层次

Location: Irvine, California, USA
Client: Orange County Great Park Corporation
Landscape Design: Ken Smith Landscape Architect

项目地点：美国加州欧文市
客　　户：橙子县大公园委员会
景观设计：肯·史密斯景观公司

Observation Balloon Preview Park Orange County Great Park

橙郡大公园之气球全景预览公园

Features 项目亮点

Taking the overall park environment into account, the design creates a strong sense of landscape preview platform which is the highest point in the entire park.

设计从整体的公园环境出发，塑造了一个纵身空间感极强的景观预览平台，是整个大公园环境中的制高点。

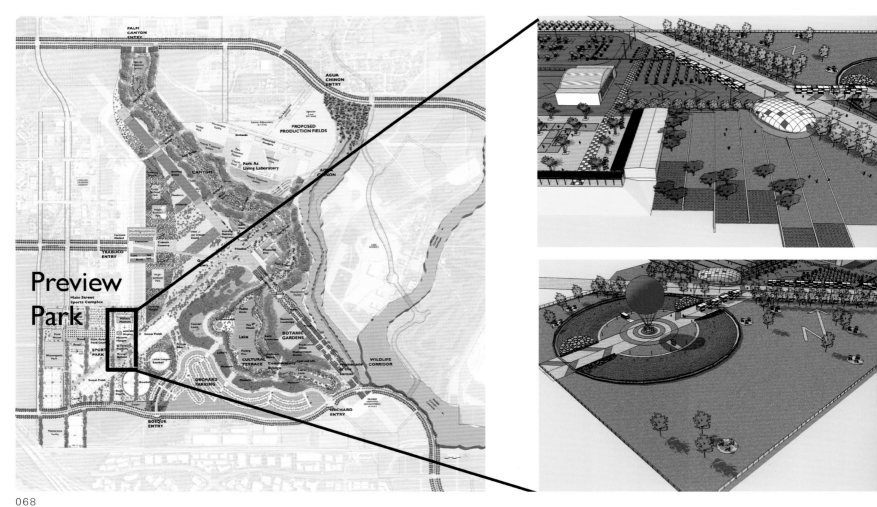

| ECOLOGY PARK 生态公园 | CULTURAL PARK 文化公园 | COMMUNITY PARK 社区公园 | CENTRAL PARK CBD 公园 |

Overview

Known regionally as "The Great Park", this 5,451,116 m² project is in El Toro Marine Air Station in Orange County, California.

This long-term project will take many years to build and grow. Part of its development plan is a 109,265 m² Preview Park that serves as a visitor center, observation area and prototyping area for elements and features being designed for the Great Park.

项目概况

在当地被称作"大公园"的南加州橙郡大公园，位于 El Toro 美国海军陆战队空军基地内，总占地 5 451 116 m²。

大公园工程将耗时数年，气球全景预览公园是整个橙郡大公园发展规划的一部分，它占地 109 265 m²，是作为大公园的游客中心和全景预览区。同时，它也是大公园设计的景观元素和景观特征的样板展示区。

LEISURE AND TOURISM LANDSCAPE 休闲度假景观

| TOURISM PARK 旅游度假公园 | CITY PARK 城市公园 |

| ECOLOGY PARK 生态公园 | CULTURAL PARK 文化公园 | COMMUNITY PARK 社区公园 | CENTRAL PARK CBD公园 |

LEISURE AND TOURISM LANDSCAPE 休闲度假景观

| TOURISM PARK 旅游度假公园 | CITY PARK 城市公园 |

| ECOLOGY PARK 生态公园 | CULTURAL PARK 文化公园 | COMMUNITY PARK 社区公园 | CENTRAL PARK CBD公园 |

LEISURE AND TOURISM LANDSCAPE 休闲度假景观

TOURISM PARK 旅游度假公园 | CITY PARK 城市公园

Keywords 关键词

- Spatial Hierarchy 多元层次
- Leisure 休闲性
- Share 共享性
- Sense of Place 空间感

Location: Bristol, UK
Client: Bristol City Council / Parkwood Leisure
Team: FoRM Associates + LA Architects

项目地点：英国布里斯托尔
客　　户：布里斯托尔议会 / Parkwood Leisure
设计团队：英国 FoRM Associates 建筑事务所 + 英国 LA Architects 建筑事务所

Hengrove Leisure Centre
Hengrove 休闲中心

Features 项目亮点

Its special feature is to merge Hengrove Park with the plaza, so as to create seasonable layout and rich spatial hierarchy for the city leisure center.

设计的特色在于将公园与广场融为一体，以化零为整的手法塑造了一个布局合理、层次丰富的城市休闲空间。

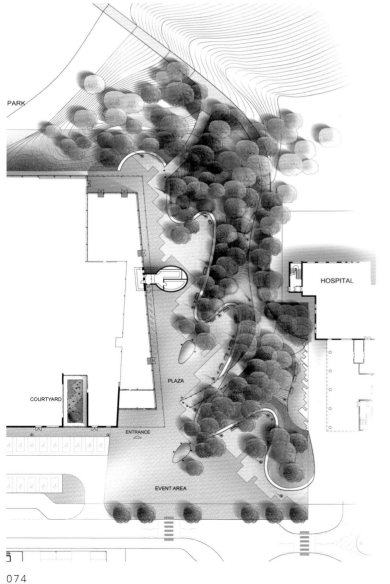

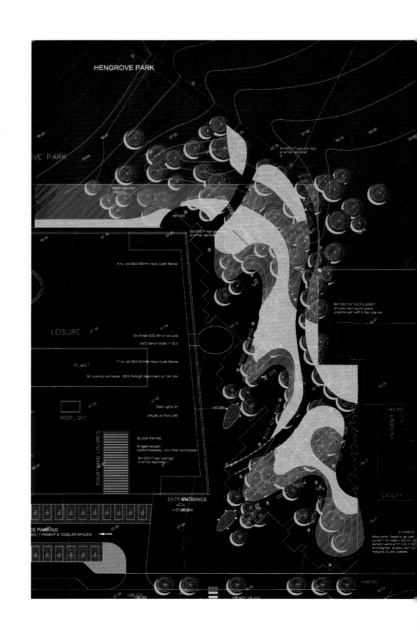

| ECOLOGY PARK 生态公园 | CULTURAL PARK 文化公园 | COMMUNITY PARK 社区公园 | CENTRAL PARK CBD 公园 |

Design Description

The central thesis involves merging Hengrove Park with the plaza and in so doing creates a spatial hierarchy that fulfils the following key ambitions:

– Creating a symbolic entry point into Hengrove Park integral with the plaza;

– Forming a distinctive landform structure as the back bone to spatial enclosure both intimate and expansive;

– Providing contoured embankments allowing DDA compliant access between the Leisure Centre and the Hospital without the use of retaining walls or steps;

– Presenting green aspects to the both the Leisure Centre and Hospital whilst creating a soft focal point and backdrop to the termination of the central boulevard;

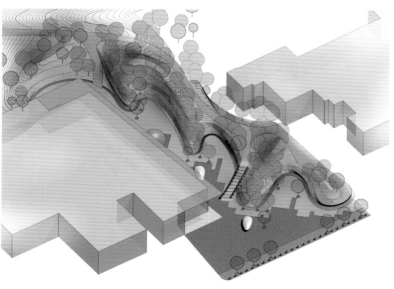
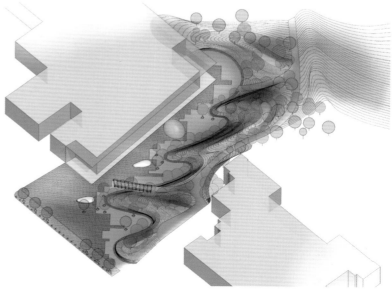

LEISURE AND TOURISM LANDSCAPE 休闲度假景观

| TOURISM PARK 旅游度假公园 | CITY PARK 城市公园 |

| ECOLOGY PARK 生态公园 | CULTURAL PARK 文化公园 | COMMUNITY PARK 社区公园 | CENTRAL PARK CBD公园 |

Delivering the component features as a catalyst to further the 'sense of place' characteristic of successful public realm domains. The overall form follows an allegorical representation of ancient wooded upland spurs meeting lowland flood plains spilling into the space between buildings; in effect drawing the park through the plaza. A series of folds provide sanctuary and intimacy culminating in a broad expanse of paving sufficient to facilitate minor local events.

设计说明

项目将 Hengrove 公园与广场合并为一体，并在空间上创造出层次感，主要体现在以下几个方面：Hengrove 公园与广场共同创造一个象征性的入口；独特的地貌结构，使各空间之间看起来联系更紧密也更宽敞；使用波浪形的路堤，休闲中心和医院之间无需挡土墙或小阶梯；休闲中心和医院之间建有大片的绿色空间，而在中央大道终点则转换为整个项目的软焦点和背景。

同时项目化零为整，进一步成功突出公共领域作为一个整体的特色"地域感"。整体的布局形式参照了古代的建筑表现形式，树木繁茂的高地与低地平原交汇于建筑物之间，有效地将公园设置在广场里面。一系列起伏的凹地，为当地人的活动空间最大限度地提供庇护和建立更亲密的关系。

LEISURE AND TOURISM LANDSCAPE 休闲度假景观

TOURISM PARK 旅游度假公园   CITY PARK 城市公园

Keywords 关键词

- Recreation 休闲属性
- Wall Structure 墙体结构
- Pavement 铺装
- Environment 环境

Location: Amsterdam, the Netherlands
Architects: Carve, Marie-Laure Hoedemakers
Carve Team: Elger Blitz, Mark van der Eng, Jasper van der Schaaf, Lucas Beukers
Area: 8,400 m²

项目地点：荷兰阿姆斯特丹
项目设计：Carve, Marie-Laure Hoedemakers
Carve 设计团队：Elger Blitz, Mark van der Eng, Jasper van der Schaaf, Lucas Beukers
面　　积：8 400 m²

Bijlmer Park
比尔梅公园

Features 项目亮点

Residential environment is brought into the redevelopment of the park. By reconfiguring the spatial and social structure, it creates a new space for recreation, entertainment and meeting.

本案在设计上将住区环境纳入了公园翻新的范围之中，通过对空间及环境的重构，为当地的市民提供了一个休闲、娱乐和集会的新空间。

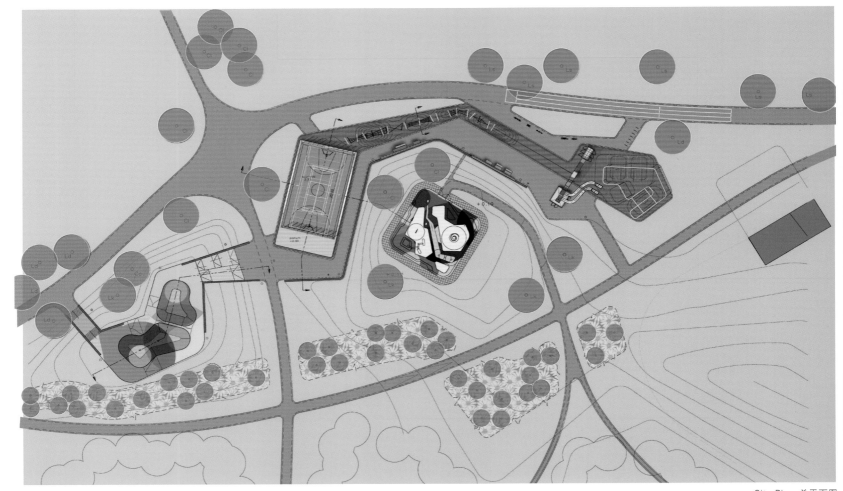

Site Plan 总平面图

| ECOLOGY PARK 生态公园 | CULTURAL PARK 文化公园 | COMMUNITY PARK 社区公园 | CENTRAL PARK CBD 公园 |

Overview

The Bijlmer Park is the main park in Amsterdam's Southeast district "Bijlmermeer". This 1960s and 1970s modernistic suburb of Amsterdam, characterized by high-rise residential and disjunctive infrastructural networks for pedestrians, cyclists and motorists where services and facilities were few and far between, had developed numerous social problems by the end of the 1980s. Radical, integral restructuring process was initiated. The renewal of the Bijlmer Park is the final chapter in this process.

项目概况

比尔梅公园是庇基莫米尔地区最主要的公园。这块位于阿姆斯特丹郊区的场地在20世纪六七十年代尚属现代化，以高层住宅楼、杂乱的交通网络及贫乏的配套设施而闻名，曾在80年代末衍生出许多社会问题，因此需要进行整体改建。比尔梅公园的翻新是阿姆斯特丹东南部城市改建工程的最后步骤。

LEISURE AND TOURISM LANDSCAPE 休闲度假景观

TOURISM PARK 旅游度假公园

CITY PARK 城市公园

Design Description

Bijlmer Park was both to remain the main park in the built-up area and was identified as a new residential environment with a program of approximately 900 dwellings. The main components are a park encircling a central sports facility and residential units along the flanks of the park. The concept reconfigures the spatial and social structure. The new housing is facing the park, providing eyes and ears on the park. The central position of the sports park keeps this facility within walking distance of the residents.

The sports and game esplanade in the center of the park is implemented as a bypass of the main route: the circular pedestrian and bicycle park-route. The esplanade embraces several elements: a ball court, the playing strip, the 'king crawler', a skate-park and a water and sand playground.

The multi-sports court includes a stage and ball-catchers with 'professional grade boulder-routes. A series of yellow frames on bright and sparkling pink safety surfacing makes the playing strip, containing different types of rope bridges and a zipline connecting to the 'king crawler'.

The 'king crawler' structure is a multilevel playing wall that incorporates facilities for the playground manager and two public toilets. The playing strip is located at the foot of two rolling green hills with trees. On top of the hill a skate-park consisting of two connected pools is hidden, with banks and stairs coming down to ground-level again. On top of the other hill one can find a water and sand playground, a colorful landscape for the youngest children with sandboxes and water-jets.

| ECOLOGY PARK 生态公园 | CULTURAL PARK 文化公园 | COMMUNITY PARK 社区公园 | CENTRAL PARK CBD公园 |

设计说明

七梅尔公园翻新工程既保留了原来的主题公园，同时还新增了可容纳约900户的住区环境。包围着中心运动设施的公园和公园两边的住宅区是项目的主要组成部分。通过空间结构及社会环境的重构，新的住宅区将面向公园，使住户能够尽享公园景观。居民只需步行就能从住宅区走到公园和游乐中心进行运动，非常方便。

位于公园中央的运动和游乐广场作为主路，是环型人行道和自行车停靠线路的分支，其中设有丰富多样的场地，如球场、玩耍地带、攀滑梯、滑板运动场以及带有沙盒和喷水设施的操场等。

多功能运动场包括一个球场和一些球挡装置，并环绕有专业等级的卵石跑道。构架于明亮、耀眼的粉红色安全地面铺装之上的一系列亮黄色的框架形成了游玩活动带状场地。框架通过不同的绳索桥和一条滑索与攀爬梯相连。

攀爬梯为一个多层次的游乐墙体结构，包括一座游乐场管理亭和两间厕所。运动带位于两座绿色小山底部。位于其中一座山顶的滑板运动场由两个相连的下沉池构成。在另一座山顶上则设有嬉水和沙盒游乐场，供小朋友开心玩耍。

| LEISURE AND TOURISM LANDSCAPE 休闲度假景观 | TOURISM PARK 旅游度假公园 | CITY PARK 城市公园 |

| ECOLOGY PARK 生态公园 | CULTURAL PARK 文化公园 | COMMUNITY PARK 社区公园 | CENTRAL PARK CBD公园 |

LEISURE AND TOURISM LANDSCAPE 休闲度假景观

| TOURISM PARK 旅游度假公园 | CITY PARK 城市公园 |

| ECOLOGY PARK 生态公园 | CULTURAL PARK 文化公园 | COMMUNITY PARK 社区公园 | CENTRAL PARK CBD公园 |

LEISURE AND TOURISM LANDSCAPE 休闲度假景观

TOURISM PARK 旅游度假公园 | CITY PARK 城市公园

Keywords 关键词

Connection 连接
Bridge Design 桥梁设计
Urban Space 城市空间
Green Corridor 绿色长廊

Location: Madrid, Spain
Client: Madrid City Government
Project Team: Burgos & Garrido Arquitectos
　　　　　　　Porras & La Casta Arquitectos
　　　　　　　Rubio & Álvarez-Sala Arquitectos
　　　　　　　West 8 Urban Design & Landscape Architecture
Area: 800,000 m²
Photography: Jeroen Musch

项目地点：西班牙马德里
客　　户：马德里市政府
项目团队：Burgos & Garrido Arquitectos
　　　　　Porras & La Casta Arquitectos
　　　　　Rubio & Álvarez-Sala Arquitectos
　　　　　West 8 Urban Design & Landscape Architecture
面　　积：800,000 m²
摄　　影：Jeroen Musch

Madrid-Río / Manzanares Lineal Park

马德里曼萨纳雷斯线性公园

Features 项目亮点

The whole park is fundamentally constituted by a trilogy of initial strategic projects including various squares, boulevards and parks and a family of new bridges which connect the urban districts along the river.

整个公园分三大策略性启动项目构成基本框架，包括各种广场、林荫大道和公园，及一系列新建的桥梁，将沿岸各城市空间有机地联系起来。

| ECOLOGY PARK 生态公园 | CULTURAL PARK 文化公园 | COMMUNITY PARK 社区公园 | CENTRAL PARK CBD公园 |

Overview

The ambitious plan by Madrid's mayor Alberto Ruiz-Gallardón to submerge a section of the M30 ring motorway immediately adjacent to the old city centre within a tunnel was realised within a single term of office. The city undertook infrastructure measures over a total length of 43 km, six of them along the banks of the River Manzanares, at a total cost of six billion Euro.

The Manzanares River is set on a 69 km long basin that starts at 2,258 m high, in the Guadarrama Mountains, and ends in the Jarama River at 527 m above sea level. The strategy is based on the conviction that the river can connect the city, with the territories north and south of the city, where the natural elements of the river basin still exist. The river becomes the door or connection between the urban interior and the territorial exterior.

项目概况

这项富有雄心的计划由马德里市长 Alberto Ruiz – Gallardón 领导，在其当届任期的时间内实现，将紧邻老城区的 M30 环城高速公路置入地下隧道。马德里市为此修建了总长超过 43 km 的基础设施，其中有 6 个主项目位于曼萨纳雷斯河沿岸，项目预算共计 60 亿欧元。

曼萨纳雷斯河坐落于一个 69 km 长的盆地，从 2 258 m 高的瓜达拉马山脉延至海拔 527 m 的哈拉玛河。项目策略是使这条河流把城市连接起来又保持城市北部和南部的范围，保留流域的自然元素，让它成为城市内部和外部之间的一个通道和连接。

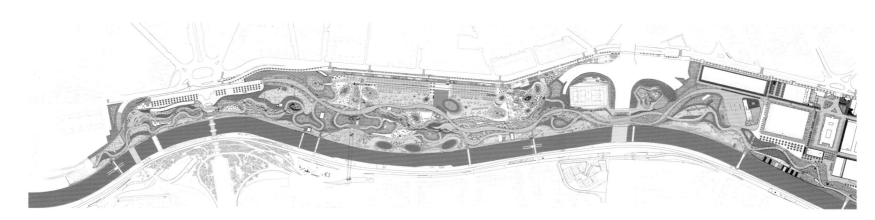

LEISURE AND TOURISM LANDSCAPE 休闲度假景观

TOURISM PARK 旅游度假公园

CITY PARK 城市公园

Design Description

The design is founded on the idea "3 + 30" – a concept which proposes dividing the 80 hectare urban development into a trilogy of initial strategic projects that establish a basic structure which then serves as a solid foundation for a number of further projects, initiated in part by the municipality as well as by private investors and residents. A total of 47 subprojects have since been developed, the most important of which include: the Salón de Pinos, Avenida de Portugal, Huerta de la Partida, Jardines de Puente de Segovia, Jardines de Puente de Toledo, Jardines de la Virgen del Puerto and the Parque de la Arganzuela. In addition to the various squares, boulevards and parks, a family of bridges were realised that improve connections between the urban districts along the river.

The Salón de Pinos, or green corridor, runs along the right river bank and links th existing and newly designed urban spaces with each other along the Manzanare River. As located almost entirely on top of the motorway tunnel, and lack o enough thick soil layers, reference to the flora of the mountains was chosen fo the outskirts of Madrid. The pine tree which is able to survive on the barren roc is planted in more than 8,000-fold. The extension of the Avenida de Portug leads towards Lisbon, in the process crosses a valley famous for its cher

| ECOLOGY PARK 生态公园 | CULTURAL PARK 文化公园 | COMMUNITY PARK 社区公园 | CENTRAL PARK CBD 公园 |

ossoms in the otherwise extremely barren and inhospitable climate of the stremadura. The abstraction of the cherry blossom as a design element of the rk, the planting of different kinds of cherry trees to extend the period in which ey flower, the reinterpretation of the Portuguese paving and the connection of e space to its surroundings have led to the creation of a popular public space.

e motive of the Huerta de la Partida has been formed with a wide variety of uit trees in groups, formed from skipping ranks. As the biggest part project with water as the dominating motive, Parque de la Arganzuela is based on the different emotions and landscapes in context of the water. The system of streams is running through the park and will form in the crossings and though the topography different spaces and motifs. Puentes Cascara is designed as a massive concrete dome with a rough texture, having the scale of park elements and not of the infrastructure. More than one hundred cables resembling whale baleens wearing the slim steel deck.

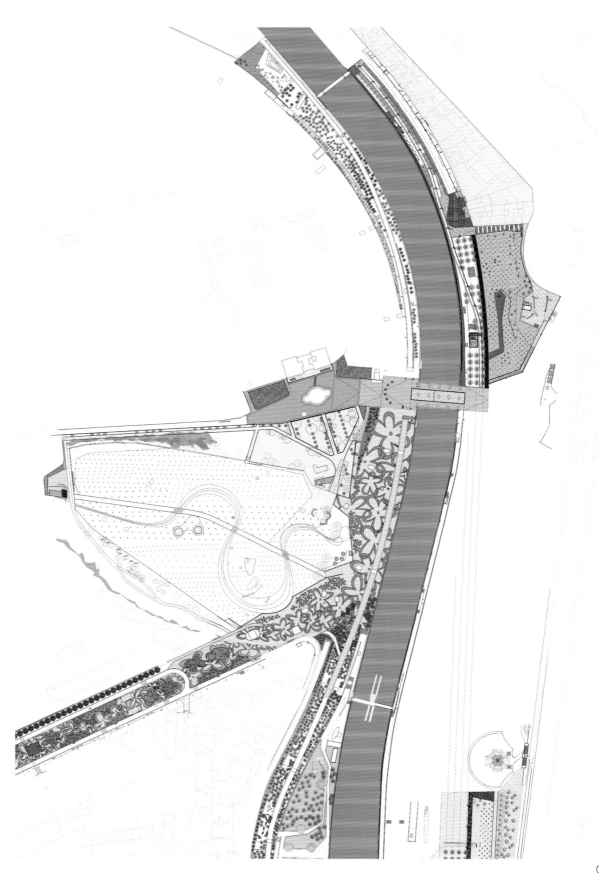

LEISURE AND TOURISM LANDSCAPE 休闲度假景观

TOURISM PARK 旅游度假公园

CITY PARK 城市公园

设计说明

设计建立在 "3 + 30" 的概念上，提出将 80 万 m² 的城市发展区分为三大策略性启动项目，这三大项目建成后形成的基本架构，可为未来大量的子项目打下坚实的基础。迄今为止，该项目已有 47 个子项目平行启动，其中最重要的包括：松林道、葡萄牙大道、果树庭园、哥维亚桥花园、托莱多桥花园、贞女桥花园、阿根苏拉公园。除了以上提到的各种广场、林荫大道和公园，一系列新建的桥梁加强了沿河各城区的联系。

松林道，即沿着右河岸的绿色长廊，连接了沿岸各个现有的和新设计的城市空间。由于几乎完全位于高速公路隧道顶部，缺乏足够厚的土层，植被方面参考了马德里郊外的山地植物。这些能够在贫瘠岩石上生长的松树种植在 8 000 多个树池中。葡萄牙大道通往里斯本的沿途会穿越一座以樱花闻名的山谷，在那里，樱花盛放于异常贫瘠荒凉的气候中。将樱花的形象抽象出来作为公园的设计元素，将不同种类的樱花树种植在樱花造型的花坛里以延长樱花短暂的花期，重新诠释葡萄牙式的传统面铺装，使之将该空间与周围环境联系起来，造就了一个人们喜爱的公共空间。

帕蒂达果树庭院的设计动机是希望通过各种各样成组栽植的果树形成不同高低朝向的对比。作为一个工程中最大的项目，阿根苏拉公园设计的主导概念是水，反映出各种与水有关的情感和景观体验。溪流流经公园，穿梭交叠，通过地形形成不同的空间和主题。卡斯卡拉桥的桥梁被设计成巨大、面粗糙的混凝土穹顶，比例尺度的运用更接近于园林小品，而不是一座城市基础设施。100 多根同鲸鱼须般的缆绳将轻薄的钢板桥面悬吊起来。

| ECOLOGY PARK 生态公园 | CULTURAL PARK 文化公园 | COMMUNITY PARK 社区公园 | CENTRAL PARK CBD 公园 |

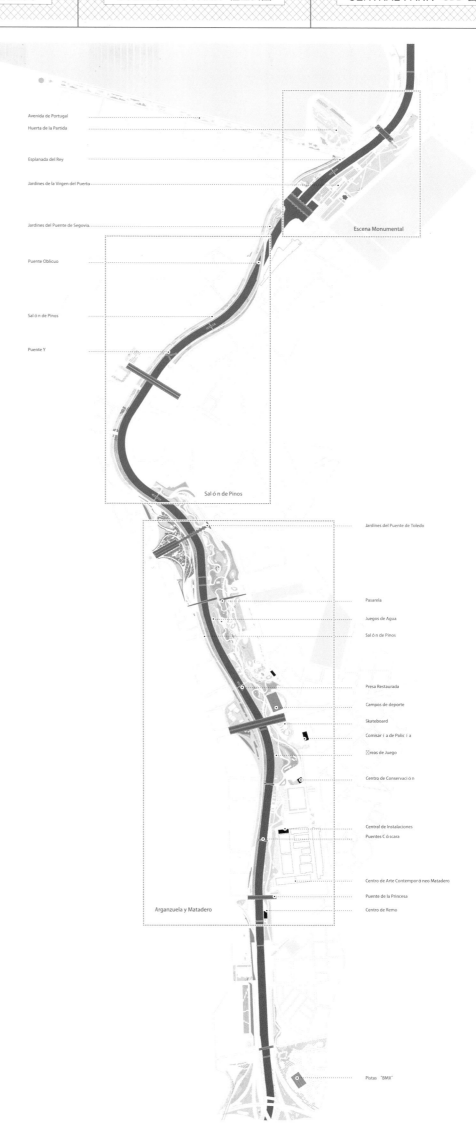

LEISURE AND TOURISM LANDSCAPE 休闲度假景观

| TOURISM PARK 旅游度假公园 | CITY PARK 城市公园 |

| ECOLOGY PARK 生态公园 | CULTURAL PARK 文化公园 | COMMUNITY PARK 社区公园 | CENTRAL PARK CBD公园 |

LEISURE AND TOURISM LANDSCAPE 休闲度假景观

| TOURISM PARK 旅游度假公园 | CITY PARK 城市公园 |

| ECOLOGY PARK 生态公园 | CULTURAL PARK 文化公园 | COMMUNITY PARK 社区公园 | CENTRAL PARK CBD 公园 |

LEISURE AND TOURISM LANDSCAPE 休闲度假景观

TOURISM PARK 旅游度假公园　　CITY PARK 城市公园

Keywords 关键词
- Color 色彩
- Theme 主题
- Comprehensive Design 综合设计
- Teach Through Activities 寓教于乐

Location: Melbourne, Australia
Client: Melbourne Municipal Government
Landscape Design: ASPECT Studio (Australia Office)

项目地点：澳大利亚墨尔本
客　　户：墨尔本政府
景观设计：澳派景观规划设计工作室

Landscape Design of Melbourne Children's Art Amusement Park

墨尔本儿童艺术游乐园景观设计

Features 项目亮点

Comprehensive landscape design methods and deliberate selection of color material, space layers, etc are applied to create a recreational space for children to learn through activities.

采用综合的景观设计手法，通过色彩、材料、空间层次等元素的精心选择，创造出独特的寓教于乐的游乐场所。

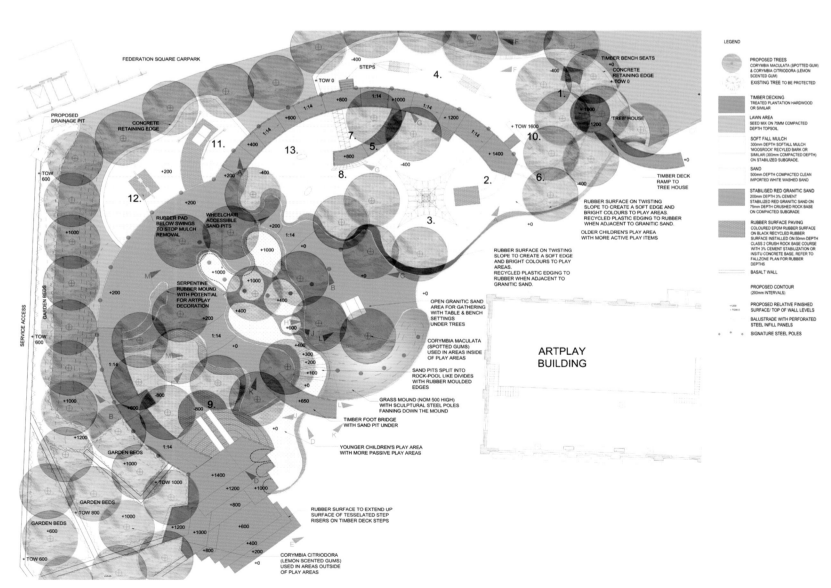

| ECOLOGY PARK 生态公园 | CULTURAL PARK 文化公园 | COMMUNITY PARK 社区公园 | CENTRAL PARK CBD 公园 |

Overview

The project is located in the city center of Melbourne, between Birrarung Marr Park and Federal Square, a place for family entertainment, therefore named "art amusement park".

The existing shape of the site is a triangle, next to a parking lot, right under the monumental sculpture of Birrarung Marr Park. The park is equipped with art office related with art amusement to offer a pleasant and unforgettable memory for Melbourne children.

项目概况

项目场地位于墨尔本市中心 Birrarung Marr 公园和联邦广场之间，是一个供家庭进行娱乐活动的场所，因此命名为"艺术游乐园"。

项目的现状是一个三角形的场地，毗邻一个停车场，在 Birrarung Marr 公园纪念雕塑下方。设有艺术游乐园相关的艺术工作室，让墨尔本的孩子们拥有一个难忘又美好的回忆。

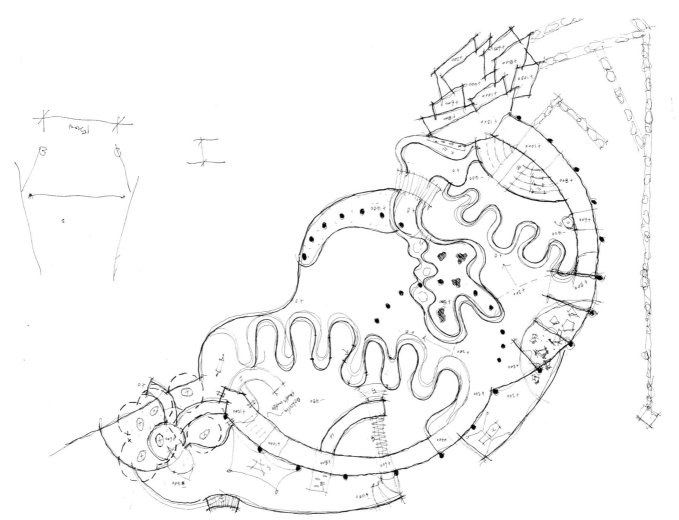

LEISURE AND TOURISM LANDSCAPE 休闲度假景观

TOURISM PARK 旅游度假公园

CITY PARK 城市公园

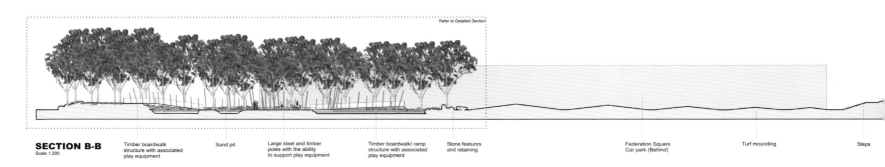

SECTION B-B
Scale 1:200

- Timber boardwalk structure with associated play equipment
- Sand pit
- Large steel and timber poles with the ability to support play equipment
- Timber boardwalk/ ramp structure with associated play equipment
- Stone features and retaining
- Federation Square Car park (Behind)
- Turf mounding
- Steps

| ECOLOGY PARK 生态公园 | CULTURAL PARK 文化公园 | COMMUNITY PARK 社区公园 | CENTRAL PARK CBD公园 |

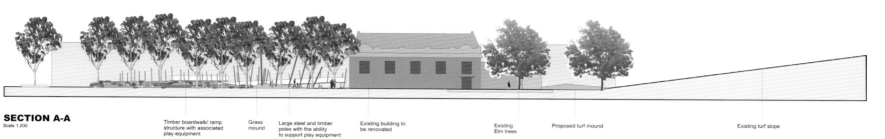

SECTION A-A
Scale 1:200

- Timber boardwalk/ramp structure with associated play equipment
- Grass mound
- Large steel and timber poles with the ability to support play equipment
- Existing building to be renovated
- Existing Elm trees
- Proposed turf mound
- Existing turf slope

LEISURE AND TOURISM LANDSCAPE 休闲度假景观

| TOURISM PARK 旅游度假公园 | CITY PARK 城市公园 |

Design Description

The purpose of this art amusement park is to provide art amusement space for children, activate their imaginations, let them enjoy in the art studio and learn knowledge through playing games. It's a unique park with a series of recreational programs. Comprehensive landscape design techniques are adopted to complete the design work from deliberate choice of materials, texture, color, gradation and space. The generous application of soil color, the smooth style and local plantation all reflect the theme of Australian Landscape and encourage people to participate more outdoor activities.

The landscape design of art amusement park gives full play of the regional advantage, integrates with the spatial edges of the circumstances, develops into a colourful amusement park and offers children joyful and surprising experience.

| ECOLOGY PARK 生态公园 | CULTURAL PARK 文化公园 | COMMUNITY PARK 社区公园 | CENTRAL PARK CBD公园 |

设计说明

艺术游乐园的目的是要为儿童提供艺术游乐场所，激发孩子们的想象力，并让孩子们在艺术工作室中游戏，寓教于乐。这是一个独特的游乐园，园内并不只是设置一系列的游乐项目，而是通过综合的景观设计手法，精心选择材料、质地、颜色、层次和空间来完成游乐园的设计。通过大手笔地采用泥土色，流畅的风格和本土植被来体现澳洲景观的主题，鼓励人们多到户外活动。

艺术游乐园的景观设计大力发挥其地段的优势，综合周围空间的优势，发展成色彩缤纷的游乐园，带给孩子们愉快、惊喜的体验。

LEISURE AND TOURISM LANDSCAPE 休闲度假景观

TOURISM PARK 旅游度假公园 | CITY PARK 城市公园

Keywords 关键词

Vitality 活力
Urban Space 城市空间
Redevelopment 重建
Landscape Network 景观网络

Location: Greenwich, London, UK
Client: Greenwich Council
Landscape Design: OKRA
In Collaboration With: Nio architecten, Buro Happold
Nature and Dimensions of Area: 19,000 m²

项目地点：英国伦敦格林威治
客　　户：格林威治市政府
景观设计：荷兰 OKRA 景观设计事务所
合作设计：Nio 建筑师事务所、Buro Happold
规　　模：19 000 m²

Cutty Sark Gardens
卡蒂萨克花园

Features 项目亮点

With functional and appearance flexibility features, it becomes an intimate and plasticity vibrant space after reconstructed by the architects.

项目最大的特色在于其灵活多变的功能和外观，通过改建，设计师将这里打造成极具亲和力和可塑性的活力空间。

| ECOLOGY PARK 生态公园 | CULTURAL PARK 文化公园 | COMMUNITY PARK 社区公园 | CENTRAL PARK CBD 公园 |

103

LEISURE AND TOURISM LANDSCAPE 休闲度假景观

TOURISM PARK 旅游度假公园

CITY PARK 城市公园

Overview

Through history the site of Cutty Sark Gardens has been a fascinating location, it represents a pioneering place with international allure. A sense of history is anchored in the area: the Meridian GMT, the Royal Naval College, and the voyages of the Cutty Sark itself. With the Olympics in mind Cutty Sark Gardens will be a gateway to other destinations, while offering an interesting focal point itself. The potential of the area is in strong contrast with the way it is presenting itself at the moment.

项目概况

卡蒂萨克花园所在位置迷人且历史悠久，它代表开创性和国际吸引力。这里富有悠久的历史意义：格林威治子午线、皇家海军学院和卡蒂萨克本身的发展历史。随着奥运会的临近，这里被赋予更加有趣的焦点的同时，还将是通往其他地方的关口。

| ECOLOGY PARK 生态公园 | CULTURAL PARK 文化公园 | COMMUNITY PARK 社区公园 | CENTRAL PARK CBD公园 |

LEISURE AND TOURISM LANDSCAPE 休闲度假景观

TOURISM PARK 旅游度假公园

CITY PARK 城市公园

Design Description

The challenges are to create an attractive place to be and to transform the area into an accessible area, which is not dominated by cars at the entrances. The neglected urban edges have to be transformed into active frontages, which contribute to the vibrancy of the area. The river walk needs to be changed into a friendly walk, as it is currently dominated by a lot of fences in the view.

OKRA suggests that the Cutty Sark Gardens will be a tidal square, reacting to flows of people: Quiet in the early hours and large amounts of visitors during holidays. People are coming, staying or passing through. The area is proposed as both urban and green, creating a gradient to the river Thames. People will experience a sequence of atmospheres: the urban network, the square around Cutty Sark, the gardens and the River Thames' Walk. The central position of the Cutty Sark will be strengthened: it will be a sculpture in between the square and the park zone.

The idea of 'tidal square' is based on functional flexibility. An inviting place which provides both intimately scaled and large scale spaces, and it has an ability of adapting to larger and smaller events. The idea is to reduce and increase space. An intimate space will be created when a wet floor and fountains reduce space and create a pool where children can play. During an event the space will be transformed when the water is removed and a large paved area is available for people to move from one place to another. A vibrant and an intimate space will be created, providing a 24h, a weekly and a seasonal rhythm. The nightscape supports this and will be both safe and interesting. Dynamic lighting and an energy efficient system provides the right amount of light during evening and night, adjusted to the amount of visitors.

Within the area a tectonic landscape provides easy access on different levels. Cutty Sark Gardens will be better connected to its urban environment by a optimized relationship between the sophisticated urban square around Cutty Sark and entrances at Greenwich Church Street and King William Walk. A cyclist and pedestrian route along the river Thames and a better connection to the Royal Naval College will improve access on the other sides.

New vibrant facades will change the relationship of Cutty Sark Gardens and the urban tissue.

Active frontages change the vitality of urban space radically. OKRA suggests architecture, which contributes to a seamless connection of the urban tissue and Cutty Sark Gardens. A gently inclining slope on the west side offers potential for commercial activities and kiosks under a green roof. In structuring the south an urban solution will be provided in a more expressive way. The built environment presents the city to the square and on the other hand creates a facade for the square and a background for the ship.

The new 'tidal square' will accommodate urban life and create green scenery that is associated with the idea of gardens and parks. In the future Cutty Sark Gardens will be a prominent area along the river Thames and to the other side by the Greenwich foot tunnel: wide views to the industrial heritage, potential of a green walk towards the Royal Naval College Green. It will be a place to linger and a place of desire.

| ECOLOGY PARK 生态公园 | CULTURAL PARK 文化公园 | COMMUNITY PARK 社区公园 | CENTRAL PARK CBD 公园 |

设计说明

计面临的挑战在于既要营造一个吸引人的地方，欢迎人们前往，同时入口处又不能停放车辆。原先被忽的城市边缘将被转变为活跃的临街界面，给整个区域注入活力。设有栅栏的河畔走道通过改建，将变得加亲切、友好。

计师将这个花园定位为"潮汐广场"，对人流具有灵活多变的功能：在凌晨显示出安静，在假期容纳大量客流量。人们在这里驻足停留，或匆匆经过。这个区域被定义为城市和绿地，创建一个到泰晤士河的过渡。们会体验到一系列的环境：城市网络，卡蒂萨克广场，花园和泰晤士河畔走道。卡蒂萨克的城市中心的位更加明显：这是广场和园区之间的一座雕塑。

"潮汐广场"的概念是基于其功能的灵活性。作为一个公共空间，既要具备亲人尺度，同时也要拥有大尺的空间，能灵活调整以适应各种大小型活动。其理念是减少和增加空间。之前的湿地空间和喷泉将被改成一个亲水池塘，供儿童玩耍。而且在活动期间，池塘的水会被引流，换上大片铺装供人们自由穿行。样一来便营造出了一个充满活力的私密空间，并可以根据24小时、每周和每个季节改变其功能。这里夜景也一样有趣，动态照明及节能系统可根据游客的多少来适时调整灯光效果，增强广场的氛围。

域内的多层次景观系统，为人们的自由通行提供了便利。花园还优化了城市广场、格林威治教堂街和国街之间的联系。沿泰晤士河岸设置了自行车道和人行走廊，另外与皇家海军学院之间也建立了通道，连了河流两岸。

的充满活力的外墙将改变卡蒂萨克花园和城市组织的关系。广场正面极具表现力和生气的地区为城市增了生机，有助于城市效区和卡蒂萨克花园的无缝连接。西侧的倾斜缓坡为将来的商业活动和绿荫报亭留了空间。而南侧的空间组织将给人留下更深刻的印象。建筑环境代表了城市广场的形象，另一方面又是场的门面和背景。

的"潮汐广场"将营造花园与公园一般的绿色景观，服务于都市生活。未来卡蒂萨克花园将成为泰晤士畔的城市亮点，纵览工业遗产，并形成一条通往皇家海军学院的绿色走廊。它将成为令人向往和流连忘的地方。

Green Architecture
Ecological Space
Natural Landscape
Sustainability

绿色建筑
生态空间
自然景观
可持续性

| LEISURE AND TOURISM LANDSCAPE 休闲度假景观 | TOURISM PARK 旅游度假公园 | CITY PARK 城市公园 |

Keywords 关键词

Farming Landscape 农业景观

Olive 橄榄树

View 视野

Material 材料

Location: Jerusalem Forest, Israel
Architect: Golany Architects
Design Team: Yaron Golany and Galit Golany
Client and Construction Team: JNF
Photography: Yaron Golany

项目地点：以色列耶路撒冷森林
建筑设计：以色列Golany建筑师事务所
设计团队：Yaron Golany, Galit Golany
客户及施工团队：JNF
摄　　影：Yaron Golany

Olive Grove, Recreation Area, Jerusalem Forest

耶路撒冷森林橄榄树园休闲区

Features 项目亮点

Re-use of an olive grove as a recreation area while retaining the original atmosphere. The contemporary interventions are sensitive and reinterpret the olive grove as a cultural legacy. The terrace walls were re-built from stones that were scattered on-site. The sitting areas were built as part of the terrace walls, to be unnoticed.

橄榄园被改造成休闲区，在保留原有气氛的条件下，重新投入使用。灵活的当代设计手法将橄榄园视为一种文化传统进行诠释。设计师现场取材，收集了周围一些散落的石块，打造了梯状挡土墙，供游客休息的桌椅安静低调地位于墙内一侧。

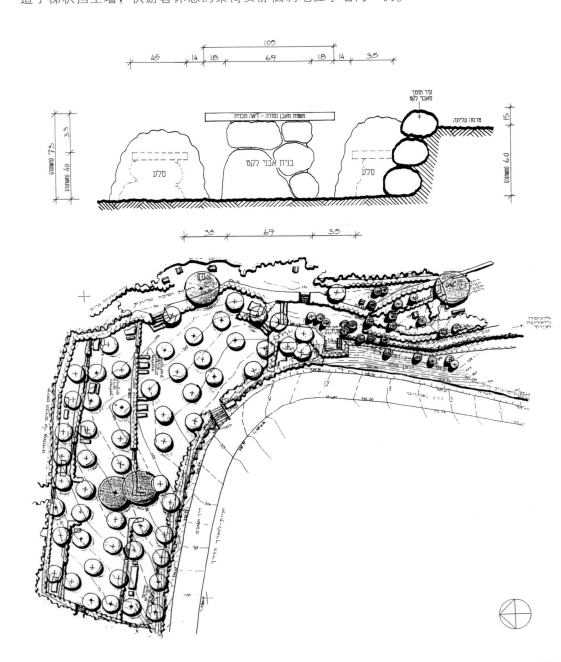

| ECOLOGY PARK 生态公园 | CULTURAL PARK 文化公园 | COMMUNITY PARK 社区公园 | CENTRAL PARK CBD公园 |

Overview

Agricultural landscapes have value beyond productivity. Olive trees are rooted in the folklore, history and culture of Israel since biblical time. Nowadays, growing population density in Israel, causes open landscape and agricultural scene to diminish in size.

项目概况

农业景观有着超越其生产力的非凡价值。橄榄树——以色列常见之景，根植于以色列的民俗、历史、文化之中。在人口激增的当今，这一开放的农业景象开始变得稀有。

LEISURE AND TOURISM LANDSCAPE 休闲度假景观

| TOURISM PARK 旅游度假公园 | CITY PARK 城市公园 |

| ECOLOGY PARK 生态公园 | CULTURAL PARK 文化公园 | COMMUNITY PARK 社区公园 | CENTRAL PARK CBD公园 |

Design Description

Therefore, when Golany Architects were commissioned to design the re-use of an olive grove for recreation, their main objective was to preserve the atmosphere and typology of the olive grove. The terrace stone walls were reconstructed from stones that were scattered on-site.

Although the project was designed referring to traditional construction methods, it does not follow any precedent, or any specific existing details. The benches and tables, were designed to blend in the terrace walls. When observed from distance they are read as part of the stone walls, without disruption to the authenticity of the grove landscape.

设计说明

当设计师接受橄榄树园的委托项目时，他们希望能最大限度地保持原有氛围。设计师现场取材，收集了周围一些散落的石块，打造了梯状挡土墙。

尽管项目的设计有涉及到传统的施工方法，但并没有一味的毫厘不差的循规蹈矩。桌椅与土墙融为一体，远远望去与树林景观毫无违和感。

LEISURE AND TOURISM LANDSCAPE 休闲度假景观 | TOURISM PARK 旅游度假公园 | CITY PARK 城市公园

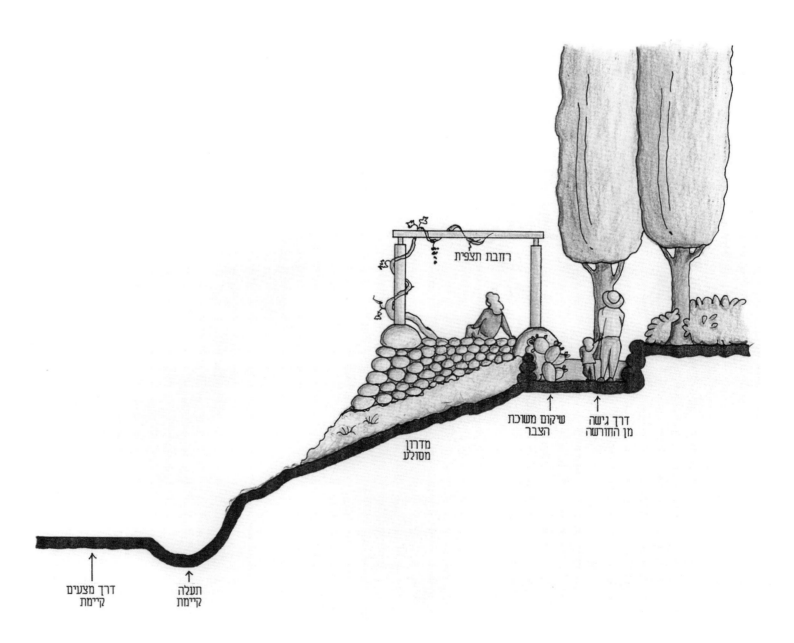

| ECOLOGY PARK 生态公园 | CULTURAL PARK 文化公园 | COMMUNITY PARK 社区公园 | CENTRAL PARK CBD 公园 |

LEISURE AND TOURISM LANDSCAPE 休闲度假景观

TOURISM PARK 旅游度假公园   CITY PARK 城市公园

Keywords 关键词

Renovation 改造重生
Natural Landscape 自然景观
Ecological 生态性
Material 材料

Location: Singapore
Client: Singapore's Public Utilities Board (PUB), National Water Agency, National Parks Board
Landscape Architecture: Atelier Dreiseitl GmbH
Area: 620,000 m², 3 km of River

项目地点：新加坡
客　户：新加坡公用事业局，国家水务局，国家公园局
景观设计：德国戴水道设计公司
面　积：620 000 m²（河道 3 km）

Bishan- Ang Mo Kio Park and Kallang River

新加坡加冷河——碧山宏茂桥公园

Features 项目亮点

The design combines renovation with eco management to create an ecological and diverse park with the reasonable arrangement of new space and the application of new materials and new technology.

设计将项目的改造重建与生态治理相结合，通过新生空间的合理布局与新材料和技术的运用打造了一个生态、多样的公园环境。

| ECOLOGY PARK 生态公园 | CULTURAL PARK 文化公园 | COMMUNITY PARK 社区公园 | CENTRAL PARK CBD 公园 |

Overview

Kallang River Bishan Park is one of the flagship projects under this programme. The park was due for major refurbishment, the Kallang River in the form of a concrete channel along the park edges was due for upgrading to cater to increased rainwater runoff from the catchment due to urbanisation. Plans were thus made to carry out redevelopment works together.

项目概况

该公园是 ABC 方案下的旗舰项目之一。由于公园需要翻新，公园旁边的加冷河混凝土渠道需要升级来满足因城市化发展造成的雨水径流排放的问题，因此这些计划被综合在一起进行此项重建工程。

LEISURE AND TOURISM LANDSCAPE 休闲度假景观

TOURISM PARK 旅游度假公园

CITY PARK 城市公园

Design Description

620,000 m² of park space have been redesigned to accommodate the inherent dynamic processes of a river system which include fluctuating water levels and widths, while providing abundant and varied recreational opportunities for park users. A 2.7 km long straight concrete drainage channel has been restored into a sinuous, natural river 3.2 km long, that meanders through the park. Three adventure playgrounds, two restaurants, a new look-out point explicitly constructed using the recycled walls of the old concrete channel, and plenty of open meadows complement the natural wonder of an ecologically restored river smack in the middle of the city. A first in the tropics, soil bioengineering techniques (a combination of vegetation, natural materials and civil engineering techniques) have been used to stabilize the river banks and prevent erosion. These techniques also create habitats for flora and fauna. The new river teem with life, and the park has already seen a 30% increase in biodiversity.

Soft, planted river banks allow people to get close to the water and in the cas of a heavy downpour, the park land that is next to the river doubles up as conveyance channel, carrying the water downstream. Bishan Park is an inspirin example of how a city park can function as ecological infrastructure, a smar combination of water source, flood management, biodiversity, recreation, an thanks to personal contact and an emotional connection with water, increasin civic responsibility towards water.

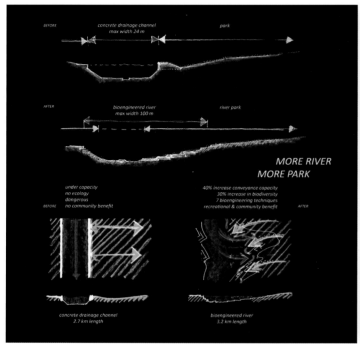

| ECOLOGY PARK 生态公园 | CULTURAL PARK 文化公园 | COMMUNITY PARK 社区公园 | CENTRAL PARK CBD公园 |

设计说明

在此项目中，62万 m² 的公园空间被重新设计，以适应河流系统固有的动态过程，其中包括水位波动和宽度，同时为公园游客提供了丰富的休闲娱乐机会。2.7 km 的加冷河从笔直僵硬的混凝土河道被改造成 3.2 km 的蜿蜒自然河道。设计师还用从旧混凝土渠道上回收利用的木材为公园建造了 3 个游乐场、2 个餐厅和一些新空间，使得城市中心拥有了充足的开放式草地和生态河流修复的自然景观。这是第一个在热带地区利用土壤生物工程技术（植被、天然材料和土木工程技术的组合）来巩固河岸和防止土壤被侵蚀的工程。通过这些技术的应用，为动植物创造了栖息地，使公园里的生物多样性增加了 30%。

公园的软景河岸使得人们更容易亲近水。同时在遇到特大暴雨时，紧挨公园的陆地，可以兼作输送通道，将水排到下游。碧山公园是一个启发性的案例，如何使城市公园作为生态基础设施，并与水资源巧妙融合起到洪水管理、增加生物多样性、提供娱乐空间等多功能。人们和水的亲密接触也增强了公民对于水的责任心。

LEISURE AND TOURISM LANDSCAPE 休闲度假景观

TOURISM PARK 旅游度假公园 | CITY PARK 城市公园

Keywords 关键词

Rainwater Collection 雨水收集

Concave Design 凹型设计

Native Plants 原生植物

Diverse Landscape 多样景观

Location: Paris, France
Client: Laguardere-Parts
Team: Agence Nicolas Michelin & Associés ANMA Architectes, Sempervirens Landscapers, Frederic-Charles AILLET, Raphaël FAVORY, Pierre SARRIEN
Area: 65,000 m²

项目地点：法国巴黎
客　户：Laguardere
设计团队：法国 Nicolas Michelin 建筑师事务所，法国 Sempervirens 景观设计，Frederic-Charles AILLET，Raphaël FAVORY，Pierre SARRIEN
面　积：65 000 m²

The Noues of Croix Catelan in Bois de Boulogne-Paris-France

法国巴黎布洛涅公园一隅

Features 项目亮点

Design from an ecological point of view, an artificial ditch is planted to collect water, forming an energy-saving green landscape system together with the diverse landscape plants.

设计从生态的角度出发，将人工沟渠作为区域内的雨水收集工具，与多样性的景观植物共同构成了一个节能、绿色的景观系统。

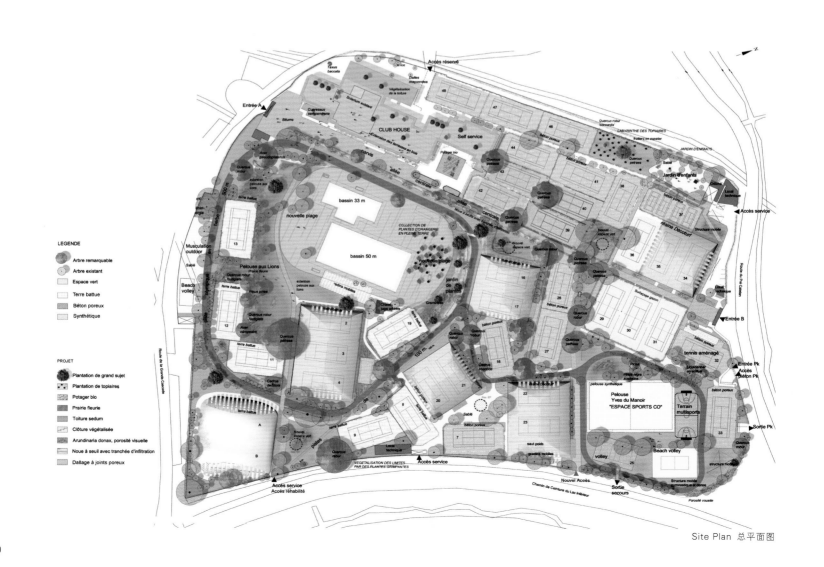

Site Plan 总平面图

| ECOLOGY PARK 生态公园 | CULTURAL PARK 文化公园 | COMMUNITY PARK 社区公园 | CENTRAL PARK CBD 公园 |

Overview

The project is located in the Bois de Boulogne covers an area of 8,460,000 m² in the west of Paris, this wood can be considered one of the "lung" of the French capital. Sempervirens Landscape worked made the landscape and environmental rehabilitation of the site in association with French Architects Nicolas Michelin ANMA. Paths have been made to remind the brick clay tennis courts, garden is structured around the theme of flowers ditches.

项目概况

项目坐落在占地 8 460 000 m² 的布洛涅森林，森林位于巴黎西部，可以说是首都绿"肺"之一。项目由法国 Sempervirens 景观设计公司与法国 Nicolas Michelin 建筑师事务所合作完成。有意打造的路径让人想起红土网球场，花园与花沟交相呼应。

| LEISURE AND TOURISM LANDSCAPE 休闲度假景观 | TOURISM PARK 旅游度假公园 | CITY PARK 城市公园 |

Design Description

A valley is planted a ditch that collects runoff. It has a double interest: Water because it restores the natural cycle of water, it prevents flooding beyond the site by storing rainwater and allowing it to infiltrate slowly. Promoting ecological native biodiversity and remarkable.

Rainwater run off from the roof of the Club House and soil and transported in stormwater ponds. Reclaimed water is used for watering the lawn. When the tanks are full, the rainwater is channeled into the valleys where water is temporarily stored and seeps into the ground. So water does not stagnate and does not allow mosquitoes to settle there.

The valleys are planted with native plant species in the Paris Basin to promote the maintenance of biodiversity of natural heritage (Typha latifolia, Equisetum arvense ...) and species chosen for their aesthetic qualities and their blue and white flowering (Meconopsis simplicifolia blue flower, white flowers Cimifuga simplex) colors of Racing Club de France).

| ECOLOGY PARK 生态公园 | CULTURAL PARK 文化公园 | COMMUNITY PARK 社区公园 | CENTRAL PARK CBD公园 |

设计说明

施工过程中，挖出一条沟渠用来收集雨水，它能促进水的自然循环、疏导强降雨，促进本地生态多样性。

屋顶雨水流经地表，最后集中到雨水池塘，用于浇灌草坪。当池塘装满后，多余的水暂时通过渠道被引进凹处或渗入地表，不会淤积，滋生蚊虫。

凹处种植着巴黎盆地的原生植物，以促进生物多样性，维护自然遗产（宽叶香蒲，马尾草……），还有一些极具美感的蓝色单叶绿绒蒿、白色单穗升麻（一蓝一白，正是巴黎赛马俱乐部的标志颜色）。

| LEISURE AND TOURISM LANDSCAPE 休闲度假景观 | TOURISM PARK 旅游度假公园 | CITY PARK 城市公园 |

Keywords 关键词

Urban Park 都市公园

Native Vegetation 原生植被

Multiple Space 多元空间

Functionality 功能性

Location: Tbilisi, Georgia
Landscape Design: CMD Ingenieros
Architects: Alberto Domingo, Carlos Lázaro, Juliane Petri
Gross Area: 50,000 m²

项目地点：格鲁吉亚第比利斯
景观设计：西班牙 CMD Ingenieros
设 计 师：Alberto Domingo, Carlos Lázaro, Juliane Petri
总 面 积：50,000 m²

Rike Park
Rike 公园

Features 项目亮点

The project pays attention to the vegetaion originality and puts differne spatial spaces into organization and combination in accordance to the exisiting nature conditions, so as to shape itself as an urban park with multiple levels.

本案在设计上强调植被的原生性，依据已有的自然条件将不同功能的空间组织串联在一起，塑造了一个层次丰富的城市公园形象。

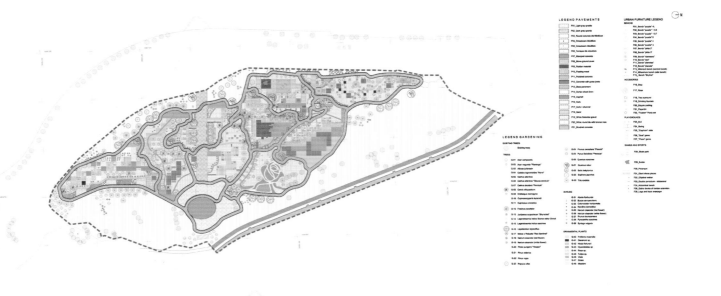

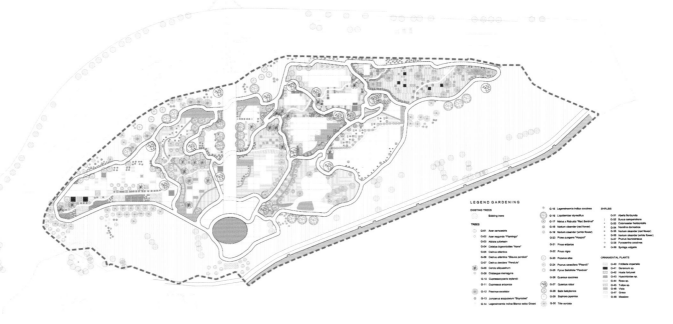

| ECOLOGY PARK 生态公园 | CULTURAL PARK 文化公园 | COMMUNITY PARK 社区公园 | CENTRAL PARK CBD公园 |

Overview

Rike Park is a new urban park in Tblisi (Georgia) designed on the Mtkvari riverside near Georgia's Presidential Palace. The project takes advantage of the pre-existing conditions: a wide space located near old thermal baths, which can be integrated in the future into the park. The landscape project uses particularly native plants and trees and the central square has a landmark – a multiple bench with the shape of a rose– referring to the Rose Revolution, an important moment on Georgia's recent history.

项目概况

项目位于格鲁吉亚总统府附近的Mtkvari河畔，是一个新的都市公园。项目充分利用已有的条件：靠近旧温泉浴场的广阔空间，未来能够将其纳入公园的范围。景观项目特意采用原生植物和树木，在中央广场设立了一个地标——一个多层次的玫瑰型长凳，意指格鲁吉亚近代史上的重要时刻玫瑰革命。

LEISURE AND TOURISM LANDSCAPE 休闲度假景观

TOURISM PARK 旅游度假公园

CITY PARK 城市公园

Design Description

The designers choose a set of local plants and trees adapting their distribution to character and uses of each zone: Eye-catching species with plenty of colour flank the entrance boulevard, inviting visitors to get to the park; picnic areas have tall and leafy trees, they guarantee sunshade during the central hours of the day in spring and summer; most-dense tree groups have been placed along the park perimeter avoiding obstacles for a global vision of the space; the specific weather in Tbilisi, colder and dryer than in other country regions, determines the type of bushes.

Each region or area plotted is defined by a main use. Picnic and playground areas are spread all around the park; there are also circuits for sport activities, meeting areas and the main square prepared for music or theatre outdoors. Oth constructions extend the program uses: a grandstand that has on the backsi the information centre; a small store designed as a chessboard. There are ve original "Sand and Water Playgrounds" with games and installations that allo children to enjoy building and modelling sand and water.

The main square and the widest streets can host temporary or permanent sta of organic food, art pieces, books, antiques, craftworks, street food carts, et Urban furniture has been designed specifically for this project considering desi and function. Many of these pieces resemble rocks or other natural elements f a better integration on the space.

| ECOLOGY PARK 生态公园 | CULTURAL PARK 文化公园 | COMMUNITY PARK 社区公园 | CENTRAL PARK CBD公园 |

设计说明

设计师根据各个区域的特点和用途选用了一系列本土植物和树木：色彩丰富而抢眼地设置在侧面入口的林荫大道，吸引游客进入园区；野餐区高大叶茂的林木在春夏能够满足遮阳的需要；最茂密的树群被安置在公园的周边，避免产生园区空间整体视觉的障碍；第比利斯与国家其他地区相比更为寒冷干燥的气候特点也决定了其灌木丛的种类。

每个区域都设有一个主要用途。野餐区和操场在整个公园随处可见，也为体育运动区、会议区和户外音乐剧院的中央广场配备了电路。在主用途模式外还延伸出了一些其他的设计：在大看台的背面设信息中心，将小商店设计成一个棋盘。有非常原始的"沙和水游乐场"，让孩子们体验用沙和水进行建筑和建模的乐趣。

主广场和最宽的街道，可以举办临时或永久的摊位，如有机食品、艺术品、书籍、古董、工艺品、街头食品车等。根据项目的景观设计和功能还配备了特意设计的城市家具，这些与岩石或其他自然元素相近的作品能够更好地融入这个空间。

LEISURE AND TOURISM LANDSCAPE 休闲度假景观

| TOURISM PARK 旅游度假公园 | CITY PARK 城市公园 |

| ECOLOGY PARK 生态公园 | CULTURAL PARK 文化公园 | COMMUNITY PARK 社区公园 | CENTRAL PARK CBD 公园 |

LEISURE AND TOURISM LANDSCAPE 休闲度假景观

TOURISM PARK 旅游度假公园

CITY PARK 城市公园

| ECOLOGY PARK 生态公园 | CULTURAL PARK 文化公园 | COMMUNITY PARK 社区公园 | CENTRAL PARK CBD公园 |

LEISURE AND TOURISM LANDSCAPE 休闲度假景观

| TOURISM PARK 旅游度假公园 | CITY PARK 城市公园 |

| ECOLOGY PARK 生态公园 | CULTURAL PARK 文化公园 | COMMUNITY PARK 社区公园 | CENTRAL PARK CBD公园 |

| LEISURE AND TOURISM LANDSCAPE 休闲度假景观 | TOURISM PARK 旅游度假公园 | CITY PARK 城市公园 |

Keywords 关键词

Combination of New and Old 新旧融合
Artificial Lake 人工湖
Voice-activated Fountain 声控喷泉
Square Space 广场空间

Location: Seoul, Korea
Client: Seoul Metropolitan City
Landscape Design: CTOPOS Design, Seoul, Korea

项目地点：韩国首尔
客　　户：首尔大都市区
景观设计：韩国首尔 CTOPOS 设计公司

West Seoul Lake Park
首尔西部湖畔公园

Features 项目亮点

The design of voice – activated fountain make the up aircraft noise utilize effectively, which not only increase the ornamental value of the park but also make local environment more Eco – friendly.

声控喷泉的设计，使位于其上的飞机噪音得以被有效地利用，不仅增加了公园景观的观赏性也使得区域环境更为生态友好。

SITE PLAN of West Seoul Lake Park

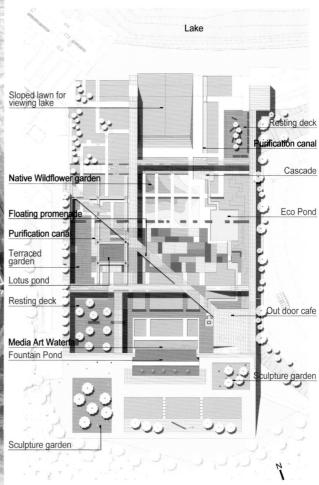

MONDRIAN PLAZA

| ECOLOGY PARK 生态公园 | CULTURAL PARK 文化公园 | COMMUNITY PARK 社区公园 | CENTRAL PARK CBD 公园 |

Overview

What is today the West Seoul Lake Park originated in 1959 as the water treatment plant. Located on the boundary between the cities of Seoul and Bucheon and reclaimed as a zone for public recreation, the park functions as meeting and communication space between the cities. The industrial facilities of the area, once infamous for having the worst living conditions in the city, were converted into an eco-friendly park in an attempt to revitalize the area and bring it up to the living standards of neighboring areas. The park was designed to mingle the themes of culture, ecology, and communication with that of rebirth.

项目概况

现今的首尔西部湖畔公园在1959年是一家水处理工厂。公园位于首尔与富川市的交界处,被改造成了一个公共游憩区,成为了两个城市之间的聚会和交流空间。该区域的工业基础设施,曾一度因被认为是城市中最差的生存环境而声名狼藉,现在却被转变成了生态友好的公园,试图借此振兴该地区,将其生活水平提高到邻近地区的标准。公园的设计将文化、生态和交流的主题与区域再生融合了起来。

LEISURE AND TOURISM LANDSCAPE 休闲度假景观

TOURISM PARK 旅游度假公园　　**CITY PARK** 城市公园

Design Description

Designers decided that a modernized and future oriented reinterpretation of existing features pursued to make the new park distinctive from others of its kind. The materials for the new park were the recycled steel pipes and concrete from the plant facility with new materials added that best matched with the old.

At the park's center, in keeping with the park's name, lies an 18,000 m² artificial lake of the kind rarely seen in downtown Seoul. The old water treatment plant is preserved and retains its natural appearance, bordered with plants and inhabited by small marine life forms. But what really attracts the eye is the sound fountain installed mid-lake, the "rebirth" of the overhead aircraft noises. The noise (when 81 dB or over) from the aircrafts coming to and from nearby Gimpo Airport sends over 15m high jets of water spewing from 41 vertical fountains that move in unison with the plane's course in the sky overhead.

| ECOLOGY PARK 生态公园 | CULTURAL PARK 文化公园 | COMMUNITY PARK 社区公园 | CENTRAL PARK CBD公园 |

Another featured space at the park is Mondrian Plaza. Old reinforced steel concrete settling tanks from the water treatment plant were torn down, leaving just a few signs of the former structures. In their place, a square garden was created introducing Mondrian structures built to a variety of scales, with horizontal and vertical lines producing a beautiful and harmonious effect. Long corten steel walls meet concrete walls in the creation of a private garden to the north and a public garden to the south, both of similar area and breadth. The square structures penetrating the corten steel walls offer charming views of tree-planted spaces. The old concrete structures blend in perpendicular harmony with new corten steel walls, forming an array of square gardens and courtyards of various sizes that intrigue visitors interested at every turn designs.

The "Media Art Fountain" is also in the Mondrian Plaza, spanning 3 m in height and 40 m in width. The design employs LED, a base for renewable energy, to offer a variety of images and music with a water playground enjoyed by flocks of child visitors for three seasons out of the year. The upper level of the settling tank structures forms 3 types of observation deck passageways, providing panoply of sequences as one overlooks the gardens and lake lying below.

LEISURE AND TOURISM LANDSCAPE 休闲度假景观

| TOURISM PARK 旅游度假公园 | CITY PARK 城市公园 |

设计说明

设计师认为,对场地现有特征的重新诠释,应追求现代化和未来化,使新公园与其他同类公园有所区别。新园区的材料是从工厂设施中回收的钢管和混凝土,以及能与旧材料较好搭配的新材料。

为与公园的名字相符,在公园的中心设置了一个在首尔市中心极为罕见的 18 000 m² 的人工湖。老污水处理厂的湖被幸运地保留了下来,并维持了其自然的外观,在其周围栽种植物,并放养小型水生生物。但这里真正的特色是安装在湖中的声控喷泉,使位于其上的飞机噪音得以被有效地利用有了新的意义。当从附近的金浦机场出发和到达的飞机所发出的噪音达到 81 分贝或以上时,就使湖中 41 个直流喷泉喷射出超过 15 m 高的水柱,并随上空飞机的飞行方向推移喷射。

公园的另一个特色空间是蒙德里安广场。污水处理厂的旧的钢筋混凝土沉淀池已被拆除,只留下

| ECOLOGY PARK 生态公园 | CULTURAL PARK 文化公园 | COMMUNITY PARK 社区公园 | CENTRAL PARK CBD公园 |

之前结构的印迹。在原场地上建起的一个方形的花园，引用了不同尺度的蒙德里安的结构，用水和垂直的线条营造出了美丽和谐的效果。长的耐候钢墙与混凝土墙共同营造了朝北的私人花园和南的公共花园，两个花园的面积和宽度相似。广场的结构渗透进了耐候钢墙中，形成了迷人的树栽植空间。旧的混凝土结构和新的耐候钢墙垂直相交，合二为一，形成一系列大小各异的方形花和庭院，每一个转角处的设计都激发着游客的勃勃兴致。

"媒体艺术喷泉"也在蒙德里安广场内，高3 m，宽40 m。设计采用了LED，以可再生能源为基础，为水上游乐场提供了各式各样的图像和音乐，成群结队的儿童游客一年中有3个季节都可以在此游玩。高处的沉淀池结构形成了3种形式的观景层通道，当人们俯瞰位于其下的花园和湖泊时，就可以看见其宏伟的序列。

LEISURE AND TOURISM LANDSCAPE 休闲度假景观

TOURISM PARK 旅游度假公园　　CITY PARK 城市公园

Keywords 关键词
- Ecological 生态
- Waterfront 滨海
- Functional 功能性
- Public Fabric 城市肌理

Location: Sydney, Australia
Client: City of Sydney Council
Landscape Design: ASPECT Studios
Photography: Florian Groehn, Adrian Boddy

项目地点：澳大利亚悉尼
客　　户：悉尼市政府
景观设计：澳派景观规划设计工作室
摄　　影：Florian Groehn, Adrian Boddy

Sydney Pirrama Waterfront Park
悉尼 Pirrama 滨海公园

Features 项目亮点

The project is designed in an ecological way that it applies the concise landscape style and material to create an ecological and charming waterfront urban space.

公园设计采用先进的生态做法，通过简洁的景观形式、材料的运用，打造一个生态、动人的滨海城市空间。

| ECOLOGY PARK 生态公园 | CULTURAL PARK 文化公园 | COMMUNITY PARK 社区公园 | CENTRAL PARK CBD公园 |

LEISURE AND TOURISM LANDSCAPE 休闲度假景观

| TOURISM PARK 旅游度假公园 | CITY PARK 城市公园 |

Overview

Overlooking the Harbour on Pyrmont Peninsula, the park combines elements of the site's history and waterfront environment with more contemporary forms and materials. The public realm includes wharfs, promenades, squares, laneways, rain gardens and a cycle way which forms significant public fabric, linking the City to the Docklands.

项目概况

站在 Pyrmont 半岛鸟瞰海港，公园的设计通过现代简洁的景观形式、材料的运用，展现场地的历史，打造出一个生态、动人的滨海城市空间。悉尼海滨公园设有海港码头、滨海大道、城市广场、小巷回廊、雨水花园、自行车道等，形成一个充满活力的城市肌理空间，增强了城市与滨海空间的联系。

| ECOLOGY PARK 生态公园 | CULTURAL PARK 文化公园 | COMMUNITY PARK 社区公园 | CENTRAL PARK CBD公园 |

Design Description

The seashore park provides Sydney citizens space for seaside leisure activity and gathering, and highlights the historic relationship between the local and the Sydney port. With the prime location in Sydney, the waterfront park creates different types of garden space to ensure its functionality and feature.

World's best practice initiatives were embedded into the master plan and rain gardens and bio-filtration trenches in the park capture and clean the water from the surrounding park storm water catchment. Street tree pits along Pirrama Rd collect street runoff and 200,000 m² water tanks have been proposed to ensure irritation is maintained sustainably throughout the year. Add to that, the proposal of solar panels on the shade canopies to power park lighting and the master plan is an exemplar of best practice ESD. Social sustainability is promoted through the creation of a significant public space at the end of Harris Street which provides an opportunity for social interaction, public gathering and displaying community wealth.

LEISURE AND TOURISM LANDSCAPE 休闲度假景观

| TOURISM PARK 旅游度假公园 | CITY PARK 城市公园 |

设计说明

海滨公园的打造为悉尼市民提供了一个滨海休闲活动与聚会的空间，并突出体现基地与悉尼海港的历史联系。由于海滨公园位于悉尼市中心的黄金地段，通过不同类型花园空间的打造，保证了海滨公园的功能性与特色。

在公园的总体规划设计中融入一系列世界上最领先、最优秀的生态做法，即在景观设计中融入雨水花园和生物过滤槽，可以在公园内收集和清洁公园集水区的水体。大道的行道树种植槽用来收集道的雨水径流，再用水箱储备 200 000 m³ 的水，保证全年的灌溉用水。此外，在公园的遮阳蓬设太阳能电池板，收集太阳能，为公园提供照明，总体规划体现了环境可持续发展的最佳范例。街道尽头建立一个重要的公共空间，有利于社区人们的互动交流、公众集会，展现社区生活美好一面。

| ECOLOGY PARK 生态公园 | CULTURAL PARK 文化公园 | COMMUNITY PARK 社区公园 | CENTRAL PARK CBD公园 |

LEISURE AND TOURISM LANDSCAPE 休闲度假景观

| TOURISM PARK 旅游度假公园 | CITY PARK 城市公园 |

| ECOLOGY PARK 生态公园 | CULTURAL PARK 文化公园 | COMMUNITY PARK 社区公园 | CENTRAL PARK CBD公园 |

LEISURE AND TOURISM LANDSCAPE 休闲度假景观

| TOURISM PARK 旅游度假公园 | CITY PARK 城市公园 |

| ECOLOGY PARK 生态公园 | CULTURAL PARK 文化公园 | COMMUNITY PARK 社区公园 | CENTRAL PARK CBD 公园 |

LEISURE AND TOURISM LANDSCAPE 休闲度假景观

TOURISM PARK 旅游度假公园　　CITY PARK 城市公园

Keywords 关键词

- Ecological 环保
- Educational 教育性
- Green Building 绿色建筑
- Natural Environment 自然环境

Location: Orange, TX, USA
Design Firm: Jeffrey Carbo Landscape Architects

项目地点：美国德克萨斯州橘子郡
设计单位：Jeffrey Carbo Landscape Architects

Shangri-La Botanical Garden

香格里拉植物园

Features 项目亮点

As Texas first LEED Platinum project, the park has become a hub of environmental awareness and education about regional landscapes and animal habitats.

该园是德克萨斯州首个获得绿色建筑白金认证的项目，已成为集环保和教育意义于一体的地方性景观区和动物栖息地。

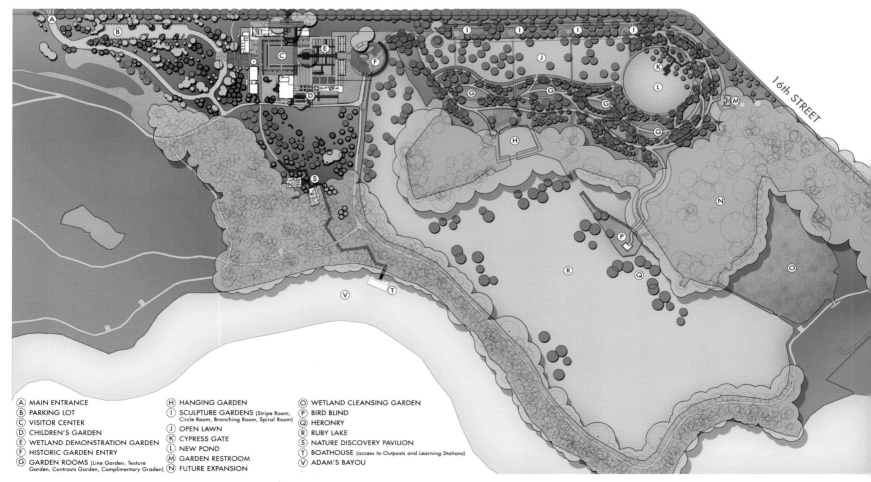

A MAIN ENTRANCE
B PARKING LOT
C VISITOR CENTER
D CHILDREN'S GARDEN
E WETLAND DEMONSTRATION GARDEN
F HISTORIC GARDEN ENTRY
G GARDEN ROOMS (Line Garden, Texture Garden, Contrasts Garden, Complimentary Garden)
H HANGING GARDEN
I SCULPTURE GARDENS (Stripe Room, Circle Room, Branching Room, Spiral Room)
J OPEN LAWN
K CYPRESS GATE
L NEW POND
M GARDEN RESTROOM
N FUTURE EXPANSION
O WETLAND CLEANSING GARDEN
P BIRD BLIND
Q HERONRY
R RUBY LAKE
S NATURE DISCOVERY PAVILION
T BOATHOUSE (access to Outposts and Learning Stations)
V ADAM'S BAYOU

Site Plan 总平面

| ECOLOGY PARK 生态公园 | CULTURAL PARK 文化公园 | COMMUNITY PARK 社区公园 | CENTRAL PARK CBD公园 |

Overview

Shangri-La Botanical Gardens and Nature Center is the first LEED Platinum-NC project in Texas. When it opened in 2008 it was one of only 50 Platinum projects in the world. The 252-acre site, in southeast, Texas, is now a hub of environmental awareness and education about regional landscapes and animal habitats. Shangri-La's design and programming make visible the life processes of many species of wildlife within the context of a native landscape, recreated botanical gardens, and innovative center for environmental education.

项目概况

香格里拉植物园与自然中心是德克萨斯州首个获得绿色建筑白金认证的项目。植物园位于德州东南部，占地约 1 019 808 m²（252 英亩）。2008 年园区对外开放时是世界仅有的 50 个白金认证的项目之一，目前它已经成为了一个集环保和教育于一体的地方性景观区和动物栖息地，能极大地激发游客的环保意识，具有深远的教育意义。在这个植物园里，许多野生动植物在自然环境中的生命发展过程都被全程呈现出来。整个园区既有创意又具有娱乐性和教育性。

LEISURE AND TOURISM LANDSCAPE 休闲度假景观

TOURISM PARK 旅游度假公园

CITY PARK 城市公园

Design Description

Its three core education zones:

• Botanical Garden: The garden was restored through a contemporary design with added layers of visual art and emphasis on design principals. The idea of line, shape, color, texture, and contrast are expressed through plant selection, patterning, and site-specific sculptures that reinforces the landscape architect's vision of a visual dialogue among native and adapted plant species, as well as art.

• Bayou/Nature Areas: Docent-led boat tours along Adam's Bayou inform visitors about the site's water ecosystem. Visitors can exit the tour at remote docks and boardwalks that lead to environmental outposts powered by solar panels.

• Birds: Blue Herons and numerous coastal bird species migrate and nest Shangri-La's Ruby Lake, thereby playing a major role in how the landscape architect selected construction methods and materials, programmed activities and staged construction.

| ECOLOGY PARK 生态公园 | CULTURAL PARK 文化公园 | COMMUNITY PARK 社区公园 | CENTRAL PARK CBD公园 |

设计说明

园内的三个核心教育区：

- 植物园：植物园设计采用现代化的设计风格，运用视觉艺术效果和强调设计原则。通过植物的选择、图案和特定地点的雕塑，以线条、形状、颜色、质地和对比度的形式强化景观设计师对当地土生植物、引进植物以及艺术的表达。

- 河口/自然区域：沿着亚当河口乘船游览，船上配有解说员介绍当地水域生态系统。游客可以在码头和浮桥处下船，那里有以太阳能电池板作电源的环保型休息亭。

- 鸟：蓝色的苍鹭和众多的海岸鸟类物种迁移和栖息在香格里拉的Ruby湖，这对景观设计师应该如何选择施工方法和材料，制定施工步骤，并分阶段施工提出了新挑战。

LEISURE AND TOURISM LANDSCAPE 休闲度假景观

| TOURISM PARK 旅游度假公园 | CITY PARK 城市公园 |

| ECOLOGY PARK 生态公园 | CULTURAL PARK 文化公园 | COMMUNITY PARK 社区公园 | CENTRAL PARK CBD公园 |

| LEISURE AND TOURISM LANDSCAPE 休闲度假景观 | TOURISM PARK 旅游度假公园 | CITY PARK 城市公园 |

Keywords 关键词
Site 场所
Passageway 通道
Rural Landscape 乡村景观
Ecological Harmony 生态和谐

Location: Yamanashi Prefecture, Japan
Landscape Design: Keikan Sekkei Tokyo Co., Ltd.
Area: +/- 7,200 m²

项目地点：日本山梨县
景观设计：(株) 景观设计 东京
面　积：约 7,200 m²

Hotarumibashi Park
Hotarumibashi 公园

Features 项目亮点

The design evolves around the unique rural landscape of the site, builds multifunctional space which fuses into surrounding environment, but als provides a space for sight view, recreation and social activities.

项目借助场地独有的乡村风景展开设计，营造出与周边环境和谐相融，可供观光、休闲、社区活动的多功能空间。

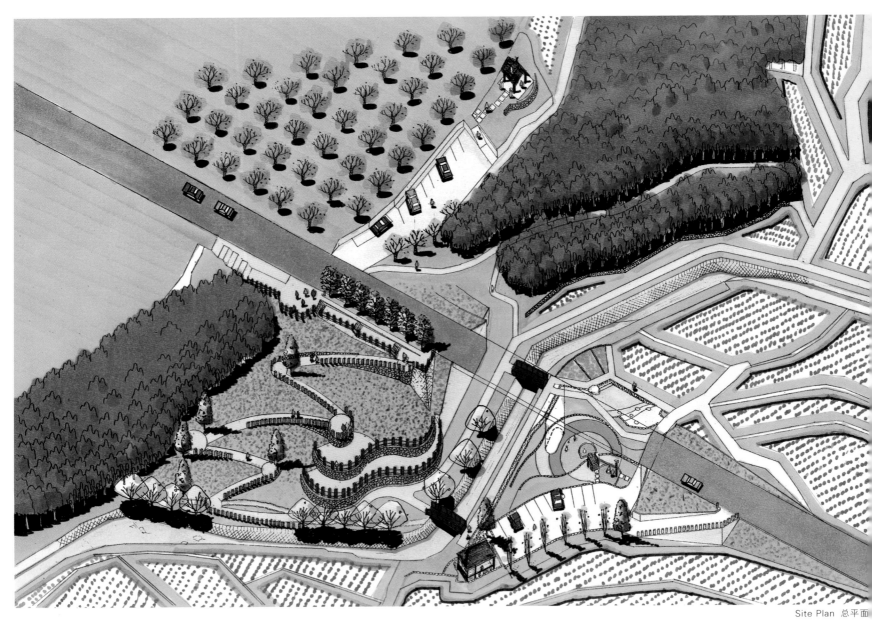

Site Plan 总平面

| ECOLOGY PARK 生态公园 | CULTURAL PARK 文化公园 | COMMUNITY PARK 社区公园 | CENTRAL PARK CBD公园 |

LEISURE AND TOURISM LANDSCAPE 休闲度假景观

| TOURISM PARK 旅游度假公园 | CITY PARK 城市公园 |

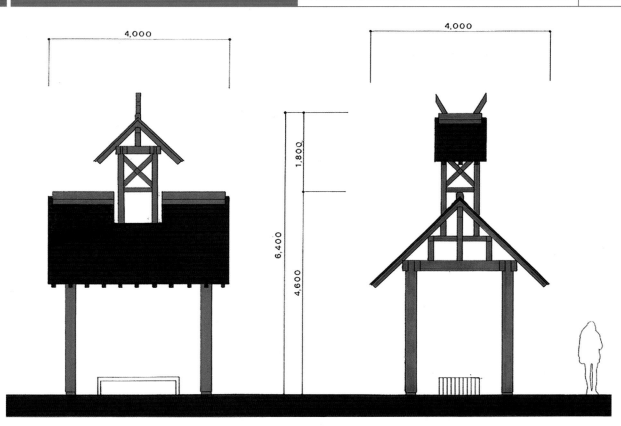

Overview

The primary objectives for the park were creating harmony with the surrounding environment, and providing a means for the community to interact and rediscover the rural landscape of the terraced rice fields, weaving together agriculture, community and ecology. Early in the 7 year process, while working with the community and government officials to refine a vision for the park, a local landowner was persuaded to donate flat land adjacent to the park site allowing for a small parking facility and the construction of a community center. The additional land area also allowed a pedestrian circulation system to be designed which connected the entire site without forcing pedestrians to cross the local roadway, which initially bisected the site.

项目概况

该公园的设计目标是打造与周边环境和谐的自然风景的同时，也为社区居民提供一处互动场所，让人们在梯田内发现淳朴的乡村风景。在七年规划过程中，通过与社区和政府部门沟通争取了一块平地作为停车场和社区住宅建筑用地。多余的空地使人行道无需穿过公园内的通道就可连接整个场地。

| ECOLOGY PARK 生态公园 | CULTURAL PARK 文化公园 | COMMUNITY PARK 社区公园 | CENTRAL PARK CBD公园 |

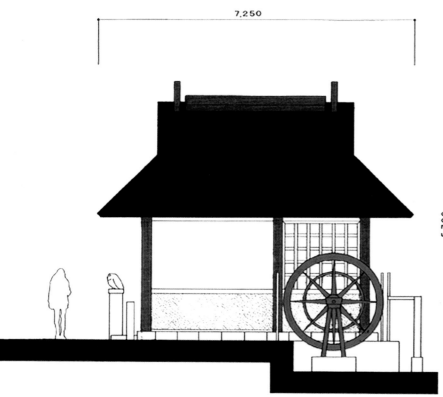
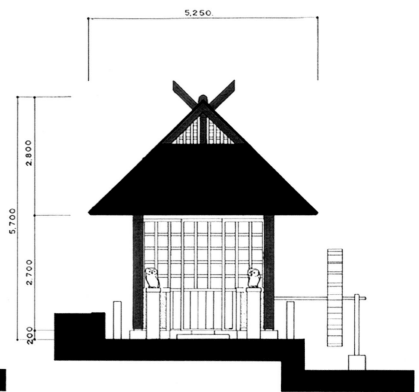

LEISURE AND TOURISM LANDSCAPE 休闲度假景观

| TOURISM PARK 旅游度假公园 | CITY PARK 城市公园 |

| ECOLOGY PARK 生态公园 | CULTURAL PARK 文化公园 | COMMUNITY PARK 社区公园 | CENTRAL PARK CBD公园 |

LEISURE AND TOURISM LANDSCAPE 休闲度假景观

| TOURISM PARK 旅游度假公园 | CITY PARK 城市公园 |

Design Description

Site elements consist of the Mount Fuji viewing plaza, terraced plaza, community center, recreated soba mill, educational facilities for schoolchildren, play areas, family picnic areas, parking, a path system, new planting, two eco-ponds, and the reconstruction of the lost habitat of the fireflies that gave the park its name (Hotarumibashi translates as "firefly viewing bridge"). The heart of the design is a two meter wide terraced plaza. Consisting of six major sections in three switchbacks, with an observation plaza at the midpoint of each switchback each section functions as a gallery utilizing shakkei, or "borrowed landscape to feature the rice terraces and hills adjacent to the park, and Mount Kushiga in the background from different angles and elevations. The result is a series images that move from a panoramic sweep of the surrounding countryside to a intimate feeling of being immersed within the landscape.

| ECOLOGY PARK 生态公园 | CULTURAL PARK 文化公园 | COMMUNITY PARK 社区公园 | CENTRAL PARK CBD公园 |

设计说明

场地特征包括富士山观光广场、梯形广场、社区中心、供娱乐用的荞麦轧机、为学生提供的教育设施、游戏区、家庭野餐区、停车场、生态池塘以及重建的萤火虫栖息地（这也是公园名字的来由——otarumibashi 可译为观赏桥的萤火虫）。设计重心是两米宽的梯形广场。三条之字形道路含六个主道，之字形道路正中央设置了观赏广场。突出梯田与四周山景之美。作为公园背景的 Kushigata 山组成了一幅展示乡村全貌的风景画，令人沉入这如画般的美景中。

LEISURE AND TOURISM LANDSCAPE 休闲度假景观 | TOURISM PARK 旅游度假公园 | CITY PARK 城市公园

| ECOLOGY PARK 生态公园 | CULTURAL PARK 文化公园 | COMMUNITY PARK 社区公园 | CENTRAL PARK CBD 公园 |

LEISURE AND TOURISM LANDSCAPE 休闲度假景观

| TOURISM PARK 旅游度假公园 | CITY PARK 城市公园 |

| ECOLOGY PARK 生态公园 | CULTURAL PARK 文化公园 | COMMUNITY PARK 社区公园 | CENTRAL PARK CBD 公园 |

LEISURE AND TOURISM LANDSCAPE 休闲度假景观

TOURISM PARK 旅游度假公园　　CITY PARK 城市公园

Keywords 关键词

Community 社区
Environment 环境
View Borrowing 借景手法
Close to Nature 亲近自然

Location: Tsuyama City, Okayama Prefecture, Japan
Landscape Design: Keikan Sekkei Tokyo Co., Ltd.
Client: Okayama Prefecture Government

项目地点：日本冈山县津山
景观设计：（株）景观设计·东京
客　　户：冈山市政府

Green Hills Tsuyama
津山绿丘

Features 项目亮点

The architects take advantage of traditional borrowed landscape, built a harmonious space with parks within sights and parks within sights, enhanced the intimate connection among city, residents and natural environment.

利用传统的日式借景手法，营造出景中有园、园中有景的和谐环境，加强了城市、居民及自然风景之间的紧密联系。

Site Plan 总平面图

| ECOLOGY PARK 生态公园 | CULTURAL PARK 文化公园 | COMMUNITY PARK 社区公园 | CENTRAL PARK CBD 公园 |

Overview

The pastoral site is located on the former grounds of the Institute of Dairy Farming of Okayama Prefecture, characterized by large expanses of rolling grassed fields spotted with small stands of mature trees and high points providing ideal views of the city below and mountain ranges surrounding the city. The intent of the design and programming of the park was to harmoniously combine the sites' potential with the existing natural features of the surrounding environment, seeking a unification of experiences between nature and man rather than contrasting a built work with its natural surroundings. The park was designed to organize the city, linking previously disconnected residential areas to each other and the surrounding landscape, providing residents an opportunity to reconnect with nature and natural process in an environment that was easily accessible and allowed for a multitude of uses.

项目概况

该项目场地为原日本冈山市乳牛业研究所，其特征是大片草地上零星地出现几棵大树，站在小丘高处可欣赏脚下优美的城市风景。该公园设计方案旨在结合场地自然特征以及周围的自然环境，追求人与自然的和谐统一。公园建成后将住宅区与周围的自然风景联系起来，也为居民提供了一处与大自然亲密接触、放松的空间。

LEISURE AND TOURISM LANDSCAPE 休闲度假景观

TOURISM PARK 旅游度假公园

CITY PARK 城市公园

Design Description

The final design preserves the "natural" landscape creating two primary use zones with the northern zone comprising a children's play field, wild flower area, and water garden which cumulatively form the "Health & Recreation" component of the park, and the southern zone forming the "Culture & Recreation" component of the park, composed of a large sloped grass lawn, an outdoor theatre, and woodland containing a music hall, museum, and artist's village. Utilizing the traditional Japanese technique of shakkei, or "borrowed landscape" the surrounding mountains were incorporated into the park creating a sense of expansiveness and connection to the natural landscape when in the park. The interplay of nature in the park and the park in nature compels visitors to explore the beauty and complexity of the natural environment increasing awareness of the natural processes surrounding them.

Green Hills Tsuyama has proven to be a valuable social and environmental asset for the community it serves, successfully providing opportunities for recreation, relaxation, and large scale cultural events, demonstrating the ability of well designed public open space to link a city and its inhabitants to each other and the surrounding natural landscape.

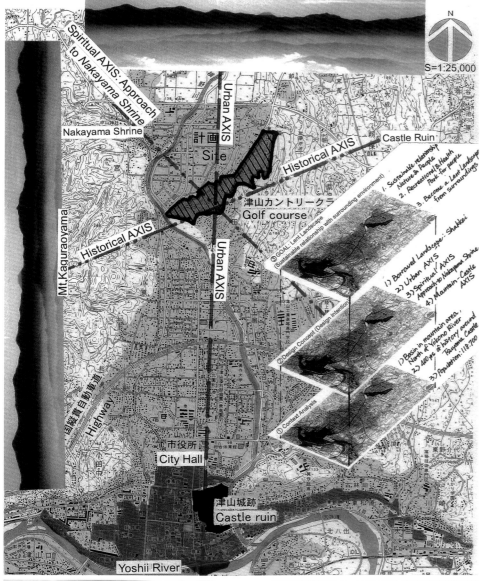

| ECOLOGY PARK 生态公园 | CULTURAL PARK 文化公园 | COMMUNITY PARK 社区公园 | CENTRAL PARK CBD公园 |

设计说明

园设计保留了自然景观，北区用地主要为儿童游戏区、野花种植区和水景花园区。南区主要作为园文化休闲娱乐设施用地，包括倾斜的草坪、户外剧院、音乐厅、博物馆以及艺术家之村。利用统的日式借景手法，周围的山景也融入公园，使公园与自然风景联系起来。景中有园，园中有景，促使游客去发掘自然环境中的美，提高人们的环保意识。

津山绿丘被认为是当地社区的社会和环境财富，为当地居民提供娱乐、放松以及文化用地，展示了公共空间作为载体联系城市、居民以及周围自然风景的重要性。

LEISURE AND TOURISM LANDSCAPE 休闲度假景观

| TOURISM PARK 旅游度假公园 | CITY PARK 城市公园 |

| ECOLOGY PARK 生态公园 | CULTURAL PARK 文化公园 | COMMUNITY PARK 社区公园 | CENTRAL PARK CBD公园 |

LEISURE AND TOURISM LANDSCAPE 休闲度假景观

| TOURISM PARK 旅游度假公园 | CITY PARK 城市公园 |

| ECOLOGY PARK 生态公园 | CULTURAL PARK 文化公园 | COMMUNITY PARK 社区公园 | CENTRAL PARK CBD 公园 |

LEISURE AND TOURISM LANDSCAPE 休闲度假景观

| TOURISM PARK 旅游度假公园 | CITY PARK 城市公园 |

| ECOLOGY PARK 生态公园 | CULTURAL PARK 文化公园 | COMMUNITY PARK 社区公园 | CENTRAL PARK CBD 公园 |

Culture
Humanities Landscape
Sense of History
Regionality

文化属性
人文景观
历史感
地域性

| LEISURE AND TOURISM LANDSCAPE 休闲度假景观 | TOURISM PARK 旅游度假公园 | CITY PARK 城市公园 |

Keywords 关键词

Sculptural Green Wall 雕塑绿墙
Native Plants 本土植物
Landscape Glass 景观玻璃
Three-Dimensional Space 立体空间

Location: Tokyo, Japan
Architects: ADPi
Clients: French Government
Landscape Design: Sempervirens Landscapers (Frédéric-Charles AILLET, Raphaël FAVORY and Pierre SARRIEN)
Constructor: Takenaka Corporation
Area: 8,000 m²

项目地点：日本东京
建筑设计：ADPi
客　　户：法国政府
景观设计：法国 Sempervirens 景观设计 (Frédéric-Charles AILLET, Raphaël FAVORY and Pierre SARRIEN)
施　　工：Takenaka Corporation
面　　积：8 000 m²

The Green Prismatic Columns / French Embassy Gardens in Tokyo

法国驻日大使馆花园

Features 项目亮点

The sculptural Green wall full of mysterious aura of art in combination with diversified native plants gave off a modern, romantic garden landscape.

The project received the Amarican Institute of Architects Prize (AIA Prize) in 2012.

雕塑绿墙的设计充满了神秘的艺术气息，通过本地植物的多样化表达，呈现了一个层次感极强的花园景观，现代、浪漫而又不拘一格。

2012 年美国建筑师协会获奖作品。

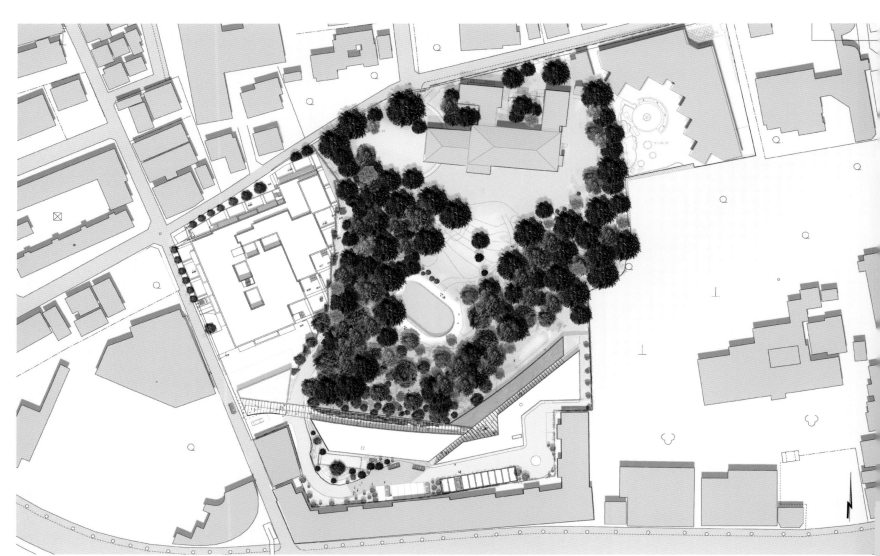

Site Plan 总平面图

| ECOLOGY PARK 生态公园 | CULTURAL PARK 文化公园 | COMMUNITY PARK 社区公园 | CENTRAL PARK CBD 公园 |

Overview

The goal of this project was to design the garden of the Embassy of France in Tokyo, Japan. Through this landscape project the designers try to respect and reinforce native biodiversity of Tokyo.

项目概况

项目的目标是打造法国驻日大使馆花园，在整个景观设计过程中，设计师试图尊重并加强东京当地的生物多样性。

LEISURE AND TOURISM LANDSCAPE 休闲度假景观

| TOURISM PARK 旅游度假公园 | CITY PARK 城市公园 |

Design Description

This sculptural Green wall is influenced by The Giant's Causeway in Ireland. The landscape project develops a real innovative concept with Japanese native plants, including sculptural organic walls where mosses artistically grow on special concrete. The wall rhythm is based on the glass frontage of the embassy.

设计说明

花园里雕塑绿墙的灵感源自爱尔兰的巨人岬。项目采用日本本土植物，如雕塑绿墙上的特殊混凝土长出的富有艺术气息的苔藓植物。使馆玻璃正立面反射着这些墙面景观。

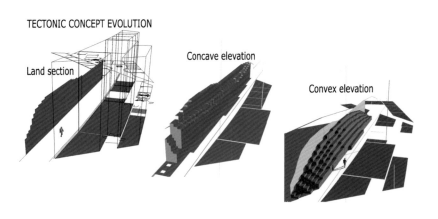
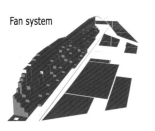
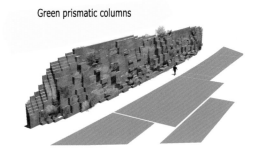
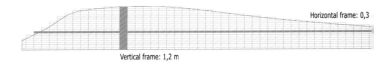
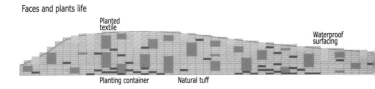
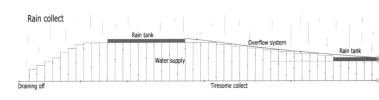

| ECOLOGY PARK 生态公园 | **CULTURAL PARK 文化公园** | COMMUNITY PARK 社区公园 | CENTRAL PARK CBD 公园 |

| LEISURE AND TOURISM LANDSCAPE 休闲度假景观 | TOURISM PARK 旅游度假公园 | CITY PARK 城市公园 |

Sharpeville Memorial Garden

沙佩维尔纪念花园

Keywords 关键词
- Recall 缅怀
- Garden Application 花园小品
- Memorial Wall 纪念墙
- Raw-Steel Poles 铁柱子

Location: Sharpeville, South Africa
Client: Urban Genesis
Design Company: GREENinc Landscape Architecture
Design Team: Anton Comrie and James French
Architect: Albonico Sack Metacity

项目地点：南非沙佩维尔
客　　户：Urban Genesis
景观设计：南非 GREENinc Landscape Architecture 景观设计事务所
设计团队：Anton Comrie and James French
建筑设计：Albonico Sack Metacity

Features 项目亮点

As a memorial garden, the project expresses people's grief feelings through the trees and bushes in the garden, and the memorial wall, amphitheatre, raw-steel poles, lawn and garden vegetation create the specific memorial atmosphere.

作为一个纪念花园的设计，项目将人们依托哀思的意念转入到花园中的一草一木中，透过纪念墙、圆形剧场、生铁柱子、草坪及花园植被等共同打造了一个特定的纪念氛围。

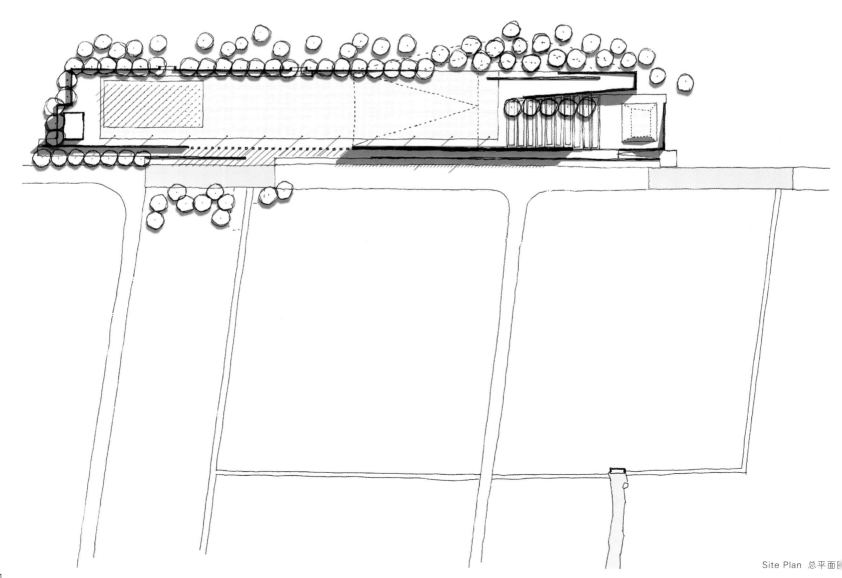

Site Plan 总平面图

| ECOLOGY PARK 生态公园 | **CULTURAL PARK 文化公园** | COMMUNITY PARK 社区公园 | CENTRAL PARK CBD 公园 |

LEISURE AND TOURISM LANDSCAPE 休闲度假景观

TOURISM PARK 旅游度假公园

CITY PARK 城市公园

Overview

The Sharpeville Memorial Garden is situated in the Phelindaba Cemetery (where the 69 graves of those killed are located) where it provides a place of remembrance and gathering for the local community. The project was conceived as a 'procession through the garden' based of the concepts of memorial, gathering and viewing. Key elements of the project are the Memorial Wall, Amphitheatre and Flowers.

项目概况

沙佩维尔纪念花园位于 Phelindaba 墓地之中。这里既是缅怀之地，也是当地社区人们集会的场所。这项工程的设计是在纪念、集会和观摩的基础上构思而成的"花园队伍"。这项工程的主要元素就是纪念墙、圆形剧场和花园。

| ECOLOGY PARK 生态公园 | CULTURAL PARK 文化公园 | COMMUNITY PARK 社区公园 | CENTRAL PARK CBD公园 |

Design Description

The memorial wall, built from clay brick, has a skeletal row of raw-steel columns along its outer edge. Each column is topped with a granite flag. These steel columns are representative of people - standing in a row, all facing the same direction. A planter in the top of the wall contains a Freylinia hedge with delicate white flowers which juxtapose the harshness of the steel and granite along the length of the wall.

Situated within the lawn space behind this wall, the 'flowers', a series of 156 unique vertical raw-steel poles each finished off with a black and white granite 'flower head', serve as a permanent bouquet of flowers laid on the memorial - akin to those left daily on graves in the cemetery.

设计说明

纪念墙是由黏土砖建造而成，在外墙的边缘上有一排骨骼型的生铁柱子。每根柱子的顶端都有一面纪念旗帜。这些钢铁柱子就像人一样，整齐地排成一排，面朝同一个方向。墙壁顶端一个花盆里面种着树篱，里面点缀着白色的小花，它与冷冰冰的钢铁和纪念旗帜共同组成了整面墙。

在墙壁后面的草坪上是花园。156根独特的垂直生铁柱子矗立在这里，顶端悬挂着黑色和白色的纪念"花环"，就像是一个永恒的花的海洋，就像每天墓地上新鲜的花朵。

LEISURE AND TOURISM LANDSCAPE 休闲度假景观

| TOURISM PARK 旅游度假公园 | CITY PARK 城市公园 |

| ECOLOGY PARK 生态公园 | CULTURAL PARK 文化公园 | COMMUNITY PARK 社区公园 | CENTRAL PARK CBD公园 |

| LEISURE AND TOURISM LANDSCAPE 休闲度假景观 | TOURISM PARK 旅游度假公园 | CITY PARK 城市公园 |

Keywords 关键词

Political Complexion 政治色彩
Cultural Characteristic 文化特性
Regionalize 地域性
Vegetation 植被

Location: Salvokop, Pretoria, Gauteng
Landscape Architect: NBGM Landscape Architecture Joint Venture (Newtown Bagale GREENinc Momo)
Architect: Office of Collaborative Architects (OCA)
Consulting Engineer: Aurecon

项目地点：南非豪登省比勒陀利亚 Salvokop 郊区
景观设计：南非 NBGM Landscape Architecture Joint Venture (Newtown Bagale GREENinc Momo) 景观设计事务所
建筑设计：Office of Collaborative Architects(OCA) 建筑事务所
工程顾问：Aurecon

The Freedom Park
南非比勒陀利亚自由公园

Features 项目亮点

It is characterized by well expressing the local culture, tradition and mental attitude through the construction of the park, and manifests the par symbolically in political form as well.

设计的特色在于通过城市公园的塑造将当地的文化、传统以及精神面貌很好地表达出来，同时又强调了公园政治层面的愿景。

| ECOLOGY PARK 生态公园 | CULTURAL PARK 文化公园 | COMMUNITY PARK 社区公园 | CENTRAL PARK CBD 公园 |

LEISURE AND TOURISM LANDSCAPE 休闲度假景观

| TOURISM PARK 旅游度假公园 | CITY PARK 城市公园 |

Overview

The Freedom Park is a project mandated by President Nelson Mandela as the natural outcome of the Truth and Reconciliation Commission process that occurred after the fall of Apartheid. Its vision is structured around four key ideas: reconciliation, nation building, freedom of people and humanity. The makin of the landscape seeks to recognise the spiritual origins of these ideas, an manifest them symbolically in physical form.

项目概况

自由公园是在南非种族隔离暴动后，由总统纳尔逊·曼德拉授权的一个真相和解委员会项目。公园有四个远景想象：和解、国民建筑、人民自由和人道主义色彩。同时通过周围园林景观的规划和实施，加强其对这些远景的精神面貌和政治色彩的渲染。

| ECOLOGY PARK 生态公园 | CULTURAL PARK 文化公园 | COMMUNITY PARK 社区公园 | CENTRAL PARK CBD公园 |

Design Description

The Freedom Park fulfils the cultural role of a Garden of Remembrance - a natural indigenous garden telling the story of South Africa's progression to freedom. It is intended as a natural symbol for reparation, a symbol of healing, a symbol of cleansing and a place where the souls of those who lost their lives in the quest for freedom can rest. It is also a place of pilgrimage, renewal and hope for all South Africans and mankind.

The Freedom Park, situated on Salvokop in Tshwane, was conceived as a narrative, a 'journey to freedom' informed by traditional African culture and Indigenous Knowledge Systems (IKS) that have not been acknowledged through past knowledge or records.

设计说明

自由公园可以说是一个回忆性的历史文化公园，可以整个地追溯南非争取自由的奋斗史。公园位于普勒陀利亚的Salvokop，园中种植有很多的植被和树木，是一个种类繁多的生物观赏园。经过不断的修复和建设，园内有很多的园林风景概念区域。

公园内还有很多公共广场以及水上喷泉区域，有时还可以看到很多具有南非特色的表演。通过这种园林景观建设，把南非的特色文化、传统以及一些精神象征传达出来。

LEISURE AND TOURISM LANDSCAPE 休闲度假景观

| TOURISM PARK 旅游度假公园 | CITY PARK 城市公园 |

| ECOLOGY PARK 生态公园 | CULTURAL PARK 文化公园 | COMMUNITY PARK 社区公园 | CENTRAL PARK CBD公园 |

195

LEISURE AND TOURISM LANDSCAPE 休闲度假景观

TOURISM PARK 旅游度假公园 | CITY PARK 城市公园

Keywords 关键词

- Natural Terraces 自然梯田
- Aromatic Plants 芳香植物
- Rural Characteristic 乡村特色
- Rotational 旋转型

Location: Pliego, Murcia, Spain.
Clients: Pliego City Council, the Ministries of Public Works and Planning, the Ministry of Presidency and Public Administration of the Region of Murcia
Design Office: Abis Architecture
Architects: Rafael Landete Pascual, Angel Benigno González Avilés, María Isabel Pérez Millán, Emilio Cortés Saura
Site Area: 21,700 m²
Photographer: David Frutos

项目地点：西班牙穆尔西亚普列戈
客　户：普列戈市议会、穆尔西亚地区的公共管理部、公共工程和规划部
设计公司：西班牙 Abis Architecture
设 计 师：Rafael Landete Pascual, Angel Benigno González Avilés, María Isabel Pérez Millán, Emilio Cortés Saura
占地面积：21 700 m²
摄　　影：大卫·弗鲁托斯

Canadas Park
Canadas 公园

Features 项目亮点

The design takes well advantage of the site characteristic on the layout of each part in the garden and skillfully implants the culture and sports activities into the natural site; the adoption of aromatic plants in the greening belt is also one characteristic of the project.

在设计上很好地利用了场地特征进行公园各部分的布局，巧妙地在自然场地中植入文化体育活动；同时绿化带中的芳香植被也是项目的特色之一。

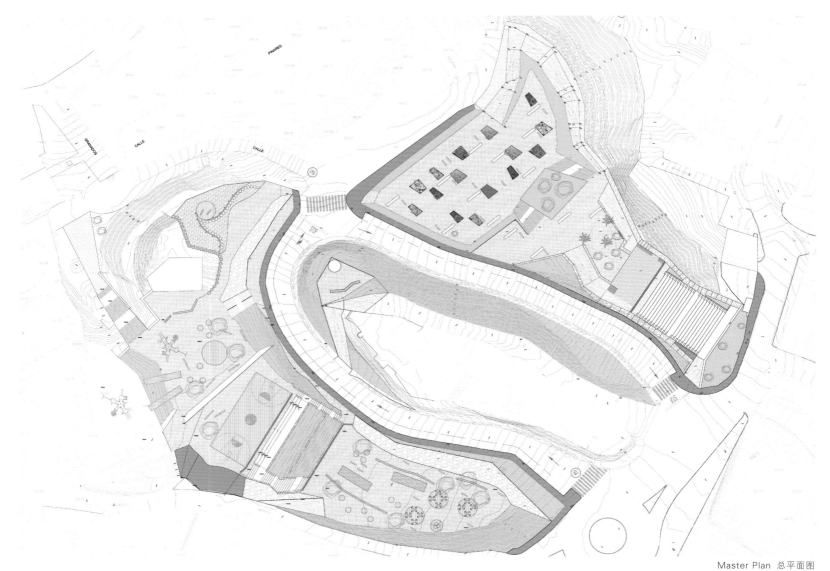

Master Plan 总平面图

| ECOLOGY PARK 生态公园 | **CULTURAL PARK 文化公园** | COMMUNITY PARK 社区公园 | CENTRAL PARK CBD公园 |

Overview

The village of Pliego is in the northwest of the province of Murcia, 42 km away from the city of Murcia. The village has an urban design characteristic of Islamic rural farmhouses. As a result the architects have an urban sinuous and disorganized, whose logic is based on its own orographic base adapting to the slope of the mountain.

The area is in the highest part of Pliego, between the urban districts of the village and the Sierra Espuña National Park. The area consists of two terraced valleys coming from the mountain. These two valleys surround a small mound which seems to move towards the urban plot, creating a beautiful view over the village with the valley at the background. The geometric duality of the site allowed the architects to implant a double program with cultural activities and sport activities exploiting the natural terraces of the initial state.

项目概况

列戈位于穆尔西亚省西北部，距离穆尔西亚市42 km，具有伊斯兰乡村农舍的城市设计特点。因此，设计师面对的是一个蜿蜒和混乱的城市，建筑物都是依地势沿山坡而建。

anadas 公园选址在 Pliego 小镇地势最高的一块地方，位于 Pliego 城区和 Sierra Espuna 国家公园之间。它包括两块沿山坡顺势而下的梯地，环绕着一个小山头，向着城镇方向移动，从山头上可以将城区景色尽收眼底。该地点的几何对偶，使设计师能够利用自然梯地的初始状态植入文化活动和体育活动双重方案。

LEISURE AND TOURISM LANDSCAPE 休闲度假景观

| TOURISM PARK 旅游度假公园 | CITY PARK 城市公园 |

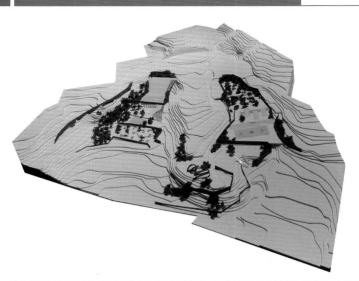

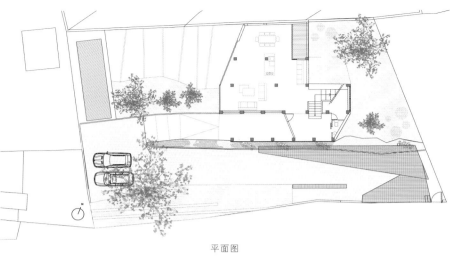

平面图

| ECOLOGY PARK 生态公园 | CULTURAL PARK 文化公园 | COMMUNITY PARK 社区公园 | CENTRAL PARK CBD 公园 |

立面剖图 1

LEISURE AND TOURISM LANDSCAPE 休闲度假景观

TOURISM PARK 旅游度假公园

CITY PARK 城市公园

Design Description

In order to integrate the spits of land on the village the architects create an urban rotational park, which generates different activities that complement Pliego and other nearby villages' leisure offer. The climate of the area becomes an essential tool for the success and the smooth running of the program, since many of the activities shall be performed outdoors. The stands naturally formed by terraces seem a gift to do an open – air amphitheater with a capacity of 1000 people.

There are different areas integrated into the park, many of them include sport and game spots, which offer a wide variety of activities for people of different age. The idea of the program and the new proposal combining the two spaces into o are interspersed with green spaces as a botanical garden with the most typic species in the area: aromatic plants (i.e. lavender, rosemary, thyme and the like All these typical species of Mediterranean forest recreate visual and olfacto sensations that permeate and surround the park reminding the visitors of t mountains of the area. These species which require inexpensive maintenan and a limited water usage will reduce the spending maintenance of the park.

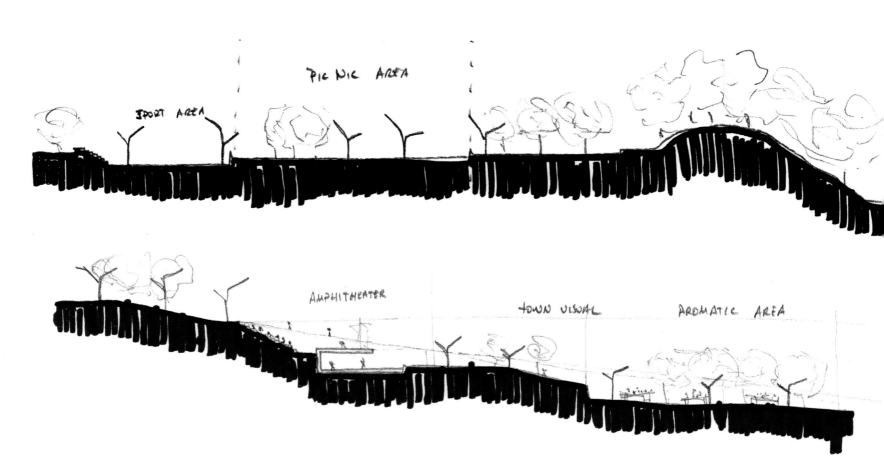

| ECOLOGY PARK 生态公园 | **CULTURAL PARK 文化公园** | COMMUNITY PARK 社区公园 | CENTRAL PARK CBD公园 |

设计说明

设计师为了整合小镇和周边的景色，设计了一个旋转型的城市公园，为游客和当地居民提供了很好的交流活动场所。这里的气候特点非常适合人们开展户外活动，因此对于设计项目的顺利完成，这一自然条件也是不可或缺的。由梯地自然形成的看台犹如自然赠予的露天剧场，可容纳1000人。

公园可细分为很多小区域，很多区域都设有运动场地和游戏场所，适合不同年龄层的人开展各种活动。文化活动区和休闲运动区之间点缀了很多绿化带，这里种植着当地特有的植被：芳香植物（薰衣草、迷迭香和百里香，等等）；这些绿化带将公园里的两大区域融为一体。这些地中海森林特有的芳香植物给人们带来了视觉和嗅觉上的美妙感受，似乎是在提醒游人，他们正身处山区。这些植物的养护成本并不高，而且需水量也不大，为公园节省了开支。

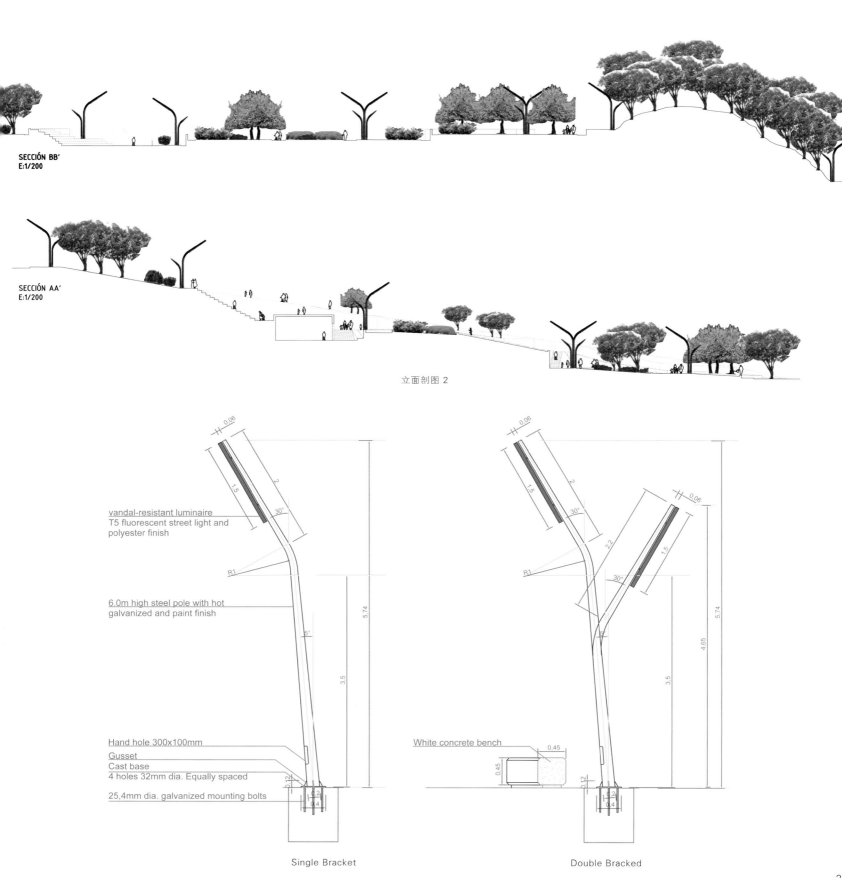

立面剖图 2

Single Bracket　　　Double Bracked

LEISURE AND TOURISM LANDSCAPE 休闲度假景观

| TOURISM PARK 旅游度假公园 | CITY PARK 城市公园 |

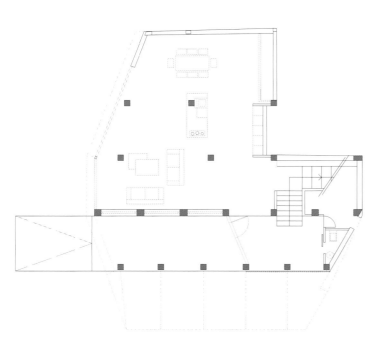
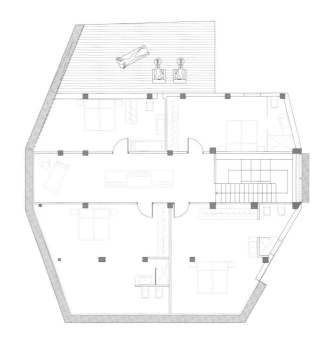

平面图

| ECOLOGY PARK 生态公园 | **CULTURAL PARK 文化公园** | COMMUNITY PARK 社区公园 | CENTRAL PARK CBD 公园 |

LEISURE AND TOURISM LANDSCAPE 休闲度假景观

| TOURISM PARK 旅游度假公园 | CITY PARK 城市公园 |

| ECOLOGY PARK 生态公园 | CULTURAL PARK 文化公园 | COMMUNITY PARK 社区公园 | CENTRAL PARK CBD 公园 |

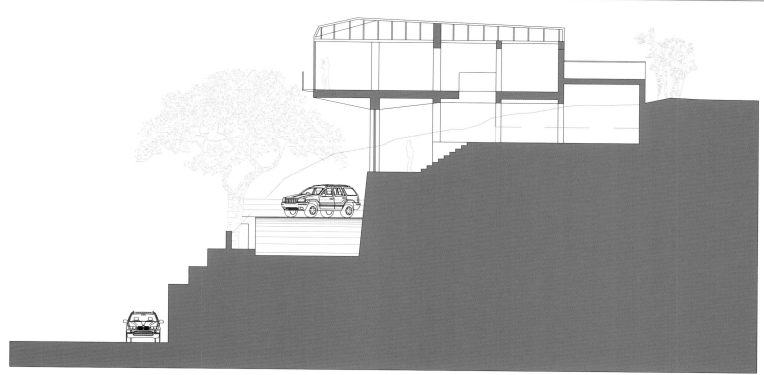

立面剖图 3

LEISURE AND TOURISM LANDSCAPE 休闲度假景观

| TOURISM PARK 旅游度假公园 | CITY PARK 城市公园 |

| ECOLOGY PARK 生态公园 | CULTURAL PARK 文化公园 | COMMUNITY PARK 社区公园 | CENTRAL PARK CBD 公园 |

LEISURE AND TOURISM LANDSCAPE 休闲度假景观

| TOURISM PARK 旅游度假公园 | CITY PARK 城市公园 |

| ECOLOGY PARK 生态公园 | **CULTURAL PARK 文化公园** | COMMUNITY PARK 社区公园 | CENTRAL PARK CBD 公园 |

| LEISURE AND TOURISM LANDSCAPE 休闲度假景观 | TOURISM PARK 旅游度假公园 | CITY PARK 城市公园 |

| ECOLOGY PARK 生态公园 | CULTURAL PARK 文化公园 | COMMUNITY PARK 社区公园 | CENTRAL PARK CBD 公园 |

| LEISURE AND TOURISM LANDSCAPE 休闲度假景观 | TOURISM PARK 旅游度假公园 | CITY PARK 城市公园 |

Würth La Rioja Museum Gardens

Würth La Rioja 博物馆花园

Keywords 关键词

Nature Vegetation 自然植被
Linear Arts 线条艺术
Ordered Space 有序空间
Leisure 休闲意境

Location: La Rioja, Spain
Landscape Design: Dom Arquitectura
Architect: Pablo Serrano Elorduy
Interior Designer: Blanca Elorduy
Collaborators: Ingeniería Torrella
Suface: 11,280 m²

项目地点：西班牙拉里奥哈
景观设计：西班牙 Dom Arquitectura 建筑事务所
设 计 师：Pablo Serrano Elorduy
室内设计：Blanca Elorduy
合作设计：Ingeniería Torrella
表 面 积：11,280 m²

Features 项目亮点

The design sets off from the plot shape and succeeds in constructing landscape space of order and balance by taking into the collocation of nature vegetation and artificial landscape.

设计从场地的形状着手，通过自然植被的配置以及人工景观的导入，成功地打造了一个有序平衡的景观空间。

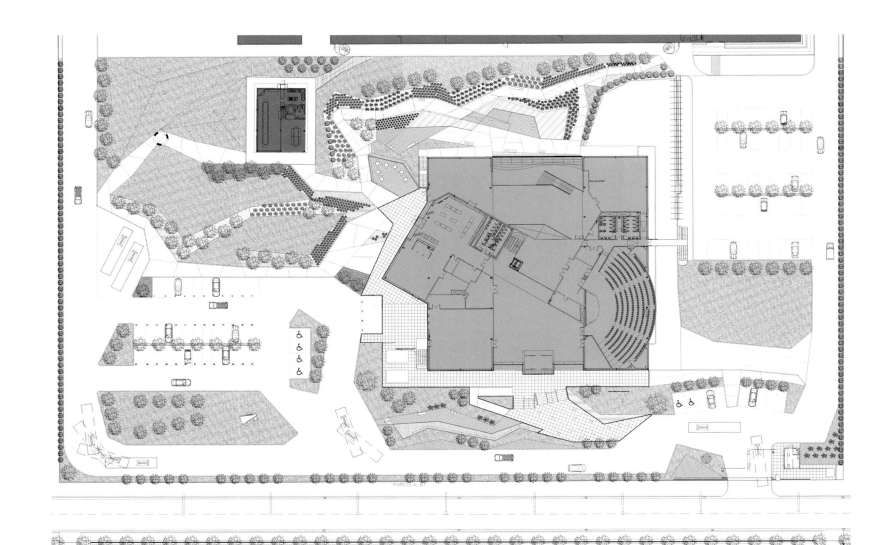

Site Plan 总平面图

| ECOLOGY PARK 生态公园 | CULTURAL PARK 文化公园 | COMMUNITY PARK 社区公园 | CENTRAL PARK CBD公园 |

Overview

The main design idea by Dom Arquitectura was to produce morphological, topographical and topological conditions and forms within the new landscape. The irregular lines are reminiscent of the nature, branches, leaves, channels, cracks, rivers. These lines seems to create a virtual grid over the existing site will geometrizing the ground, and organizing it too. So the designers managed to control each different areas of the future landscape. The result, is not an arbitrary garden, is a space where everything is in place and has a rationale balance.

项目概况

该项目的主要出发点是为新景观制造提供形态、地形、地质上的条件和形式。场地不规则的线条让人们联想到了自然、枝叶、河流和裂隙。这些线条看似在地面勾勒出了一个虚拟的网格，呈有序的几何图形。设计师最终成功地将其导入其项目景观的各个不同区域中，打造出一个有序平衡的空间，而不是一个随意的花园。

| LEISURE AND TOURISM LANDSCAPE 休闲度假景观 | TOURISM PARK 旅游度假公园 | CITY PARK 城市公园 |

Design Description

Volumes wood protruding from the ground, concrete walls and walkway like rocks that are breaking into the stillness of the vegetation. They become walking tours and rest areas. Other strips become vertical trees walls, lace permeable branches laces, bark areas reminiscent of the forest ground, white pebbles stones eroded over time by water, as if a river were involved. The concrete walk lead us for a surprising ride, where everything is moving at the same pace, herbs, trees, rocks and water sheets.

设计说明

自然植被的"宁静"被"闯入"的混凝土、原木和卵石打破，这些元素被打造成一个休闲和散步的场所。垂直的树墙、枝叶的修饰和树皮都将人们带入森林的意境，白色的鹅卵石随着时间的推移逐渐被水侵蚀，犹如一条河流曾经存在的见证。混凝土小路带人们开展一段惊喜的旅程，花草、树木、石头、水流——所有这些都随着同样的步调和速度在移动。

| ECOLOGY PARK 生态公园 | **CULTURAL PARK 文化公园** | COMMUNITY PARK 社区公园 | CENTRAL PARK CBD公园 |

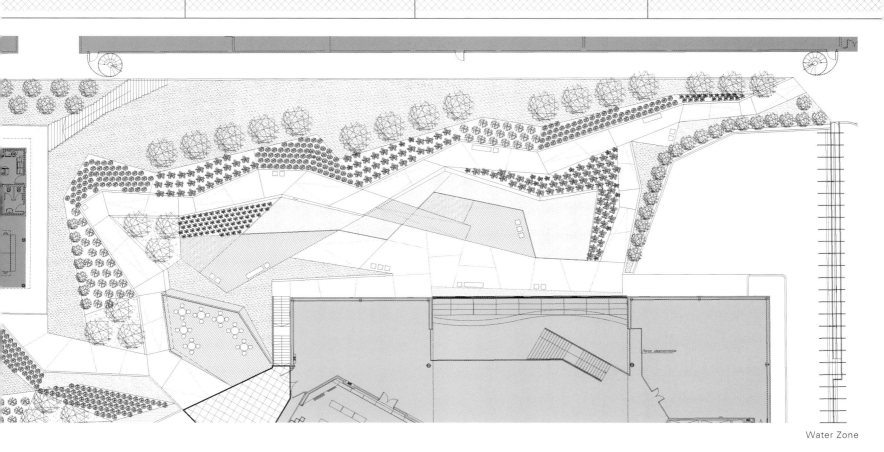

Water Zone

LEISURE AND TOURISM LANDSCAPE 休闲度假景观

| TOURISM PARK 旅游度假公园 | CITY PARK 城市公园 |

| ECOLOGY PARK 生态公园 | **CULTURAL PARK 文化公园** | COMMUNITY PARK 社区公园 | CENTRAL PARK CBD公园 |

| LEISURE AND TOURISM LANDSCAPE 休闲度假景观 | TOURISM PARK 旅游度假公园 | CITY PARK 城市公园 |

Keywords 关键词

Sculpture Park 雕塑公园

Waterfront Characteristics 滨水特征

Green Platform 绿色平台

Master Plan 整体性

Location: Seattle, Washington, USA
Landscape Design: WEISS MANFREDI

项目地点：美国华盛顿州西雅图市
景观设计：WEISS MANFREDI

Olympic Sculpture Park

奥林匹克雕塑公园

Features 项目亮点

By using an ongoing "Z" shaped green platform to put together those three parts, the design well solves the transition and connection between the new waterfront district and the city center.

设计采用一个不间断的"Z"字形绿色平台，将三部分连成一体，很好地处理了市中心与滨水新区的过渡与衔接。

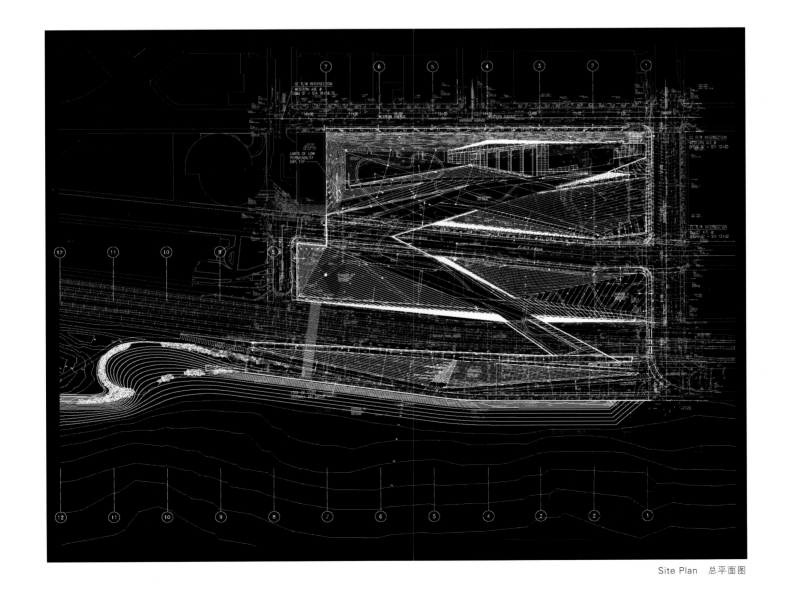

Site Plan 总平面图

| ECOLOGY PARK 生态公园 | CULTURAL PARK 文化公园 | COMMUNITY PARK 社区公园 | CENTRAL PARK CBD公园 |

Overview

Envisioned as a new urban model for sculpture parks, this project is located on Seattle's last undeveloped waterfront property - an industrial brownfield site sliced by train tracks and an arterial roads.

项目概况

作为城市雕塑公园的一个新的构思模式，场地位于西雅图最后一块未开发的滨水地区，是一片被铁轨和一条干道公路分割的工业用地。

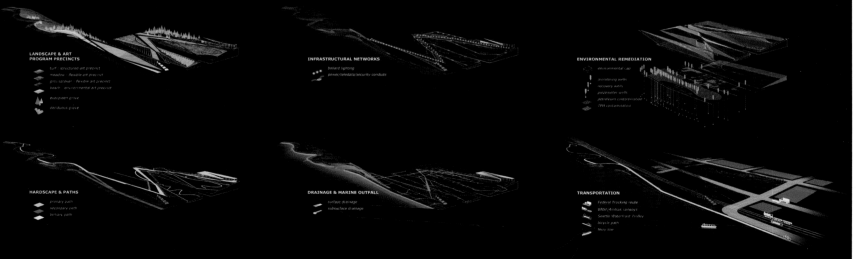

LEISURE AND TOURISM LANDSCAPE 休闲度假景观

| TOURISM PARK 旅游度假公园 | CITY PARK 城市公园 |

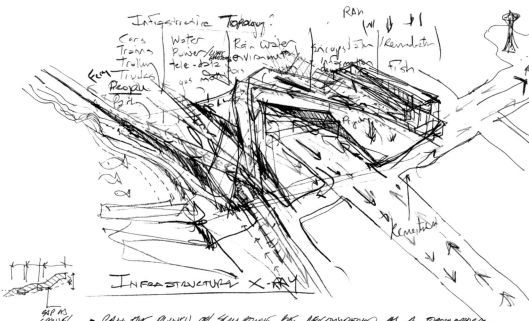

| ECOLOGY PARK 生态公园 | **CULTURAL PARK 文化公园** | COMMUNITY PARK 社区公园 | CENTRAL PARK CBD公园 |

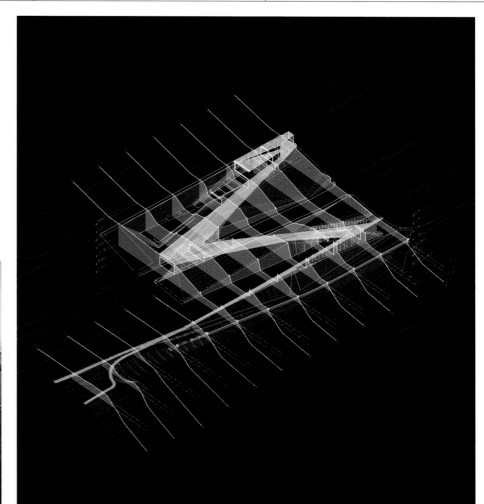

LEISURE AND TOURISM LANDSCAPE 休闲度假景观

| TOURISM PARK 旅游度假公园 | CITY PARK 城市公园 |

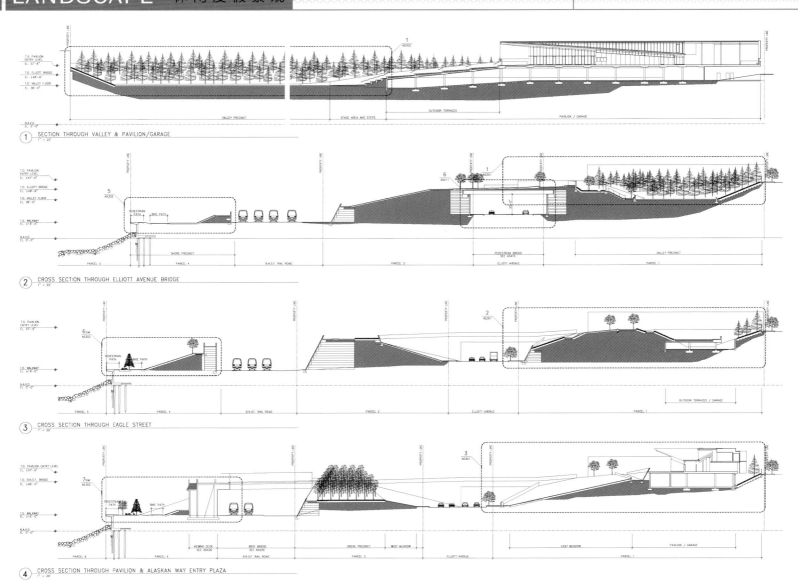

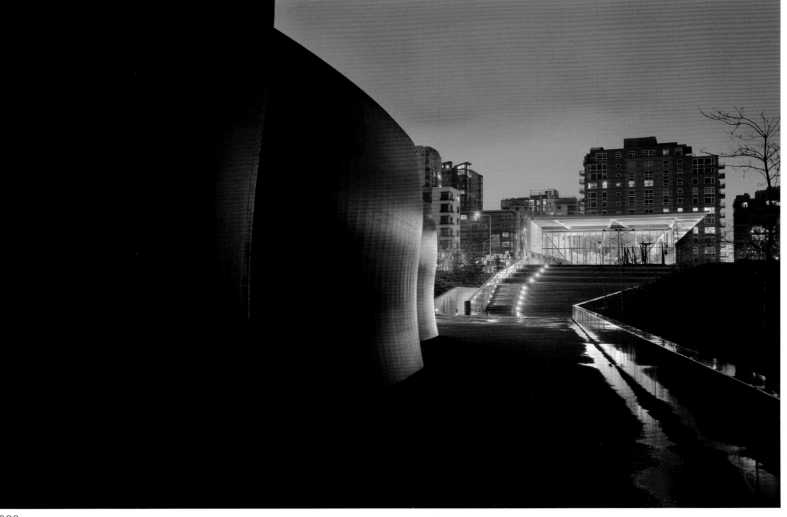

| ECOLOGY PARK 生态公园 | **CULTURAL PARK 文化公园** | COMMUNITY PARK 社区公园 | CENTRAL PARK CBD公园 |

Design Description

The design connects three separate sites with an uninterrupted Z-shaped "green" platform, descending 12 m from the city to the water, capitalizing on views of the skyline and Elliott Bay and rising over existing infrastructure to reconnect the urban core to the revitalized waterfront.

设计说明

在设计中采用一个不间断的"Z"字形"绿色"平台,将三部分连成一体。这块"绿色"平台,从市区向水域逐渐下降约12 m,利用天际线和艾略特海湾的天然景观,开发现有的基础资源,将市中心连接到复兴的滨水地区。

LEISURE AND TOURISM LANDSCAPE 休闲度假景观

| TOURISM PARK 旅游度假公园 | CITY PARK 城市公园 |

| ECOLOGY PARK 生态公园 | CULTURAL PARK 文化公园 | COMMUNITY PARK 社区公园 | CENTRAL PARK CBD 公园 |

Community Park

社区公园

Leisure
Community Environment
Functionality
Privacy

休闲属性
社区环境
功能性
私密性

LEISURE AND TOURISM LANDSCAPE 休闲度假景观

TOURISM PARK 旅游度假公园 | CITY PARK 城市公园

Keywords 关键词
Public Space 公共空间
Entertainment 娱乐
Waterscape 水源景观
Natural Purification 自然净化

Park of Luna
卢那公园

Location: Heerhugowaard-South, the Netherlands
Client: Municipality of Heerhugowaard and HAL-board
Design: HOSPER
Partners: DRFTWD Office Associates, Nelen en Schuurmans, Architectural Office Schulze en van Dijk, Sander Douma Architects
Masterplan "City of the Sun": KuiperCompagnons
Area: 1,700,000 m²
Photography: Pieter Kers, Amsterdam/Aerophoto Schiphol BV/Jan Tuijp

项目地点：荷兰南海尔许霍瓦德市
客　户：海尔许霍瓦德市政府和HAL董事会
景观设计：HOSPER景观设计事务所
合作设计：DRFTWD事务所、Nelen en Schuurmans事务所、Schulze en van Dijk建筑事务所、Sander Douma建筑事务所
太阳城总体规划：高柏伙伴公司
面　积：1 700 000 m²
摄　影：Pieter Kers、Aerophoto Schiphol BV、Jan Tuijp

Features 项目亮点

The most important feature of the design is the waterscape will be brought into the plan and separate the residential area from the surrounding recreational areas and guarantee that a 'ecology and open space' can be experienced in the plan area.

设计最大的特点在于将水源景观引入整个大环境中，将居住区与娱乐区分离，提供了一个开放、生态且具有体验性的城市休闲空间。

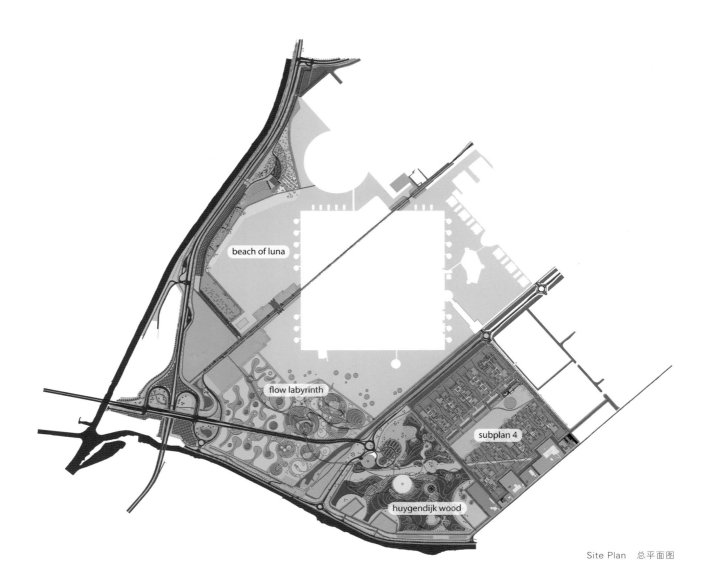

Site Plan 总平面图

| ECOLOGY PARK 生态公园 | CULTURAL PARK 文化公园 | **COMMUNITY PARK 社区公园** | CENTRAL PARK CBD公园 |

Overview

HOSPER has been working on the Park of Luna for the past ten years, from the masterplan phase up until implementation. In collaboration with the municipality, Neelen and Schuurmans and DRFTWD this resulted in the development of an attractive recreational area with several activities and a naturally purified swimming lake as it's central element. The Park of Luna has been nominated for the Rosa Barba European Landscape Prize 2010 and is, as part of the project 'City of the Sun', selected for the European Urban and Regional Planning Achievement Awards 'Special Merit Award' 2010.

Over the course of the years the traditional agrarian polder landscape south of Heerhugowaard has changed into a modern city landscape in which homes, recreation and nature development are closely interwoven with unique water protection system.

项目概况

在过去十年里，HOSPER一直致力于卢那公园的总规划和规划的实行。在市政府、Neelen and Schuurmans及DRFTWD事务所的协助下，共同完成了一个极具魅力的娱乐场所，其中包括几个活动场所，以及一个作为中心元素的可自然净化的游泳湖。卢那公园获2010罗莎·芭芭欧洲景观奖提名，并且作为"太阳城"项目的一部分，入选2010欧洲城市与地区规划成就奖的"特殊贡献奖"。

过去的几年里，海尔许霍瓦德南部传统的圩田乡村景观经过岁月变更，现已成为现代化的城市景观——住宅区、休闲区、自然开发区与独特的蓄水保护系统相互交织。

LEISURE AND TOURISM LANDSCAPE 休闲度假景观

TOURISM PARK 旅游度假公园

CITY PARK 城市公园

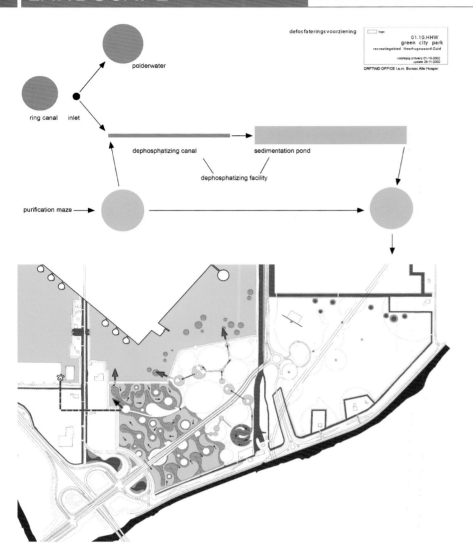

Design Description

The Stad van de Zon is completely surrounded by a 'ring of open water' which encompasses more than 70 hectares of new water. This 'ring of open water' will separate the residential area from the surrounding recreational areas and guarantee that a 'large amount of open space' can be experienced in the plan area. Most use of the water will be made of the banks in the recreational area. The recreational area has two sides: the inner side is oriented towards the open water and the Stad van de Zon, the outer side towards the surrounding landscape. The ambitious water system is unique: it is designed to store a great deal of water and conserve water in the summer. A great deal of attention was devoted to the water quality, accessibility, and the ability to experience the system. To this end a number of structures were designed which include a circulation pumping station, a natural purification plant, a dephosphatising pond, a bridge, and a canoe crossing.

Structures enable passersby to experience the water purification. The structures were designed to ensure that passersby could experience the water to the maximum possible extent. Pursuant to this objective the public can access the roof of the pumping station. This offers a view of the lake. In addition, the pumping station is located in a strategic position close to the entrance to the beach. The inlet of the water purification plant has been raised above water level where it is both visible and audible. In adopting this approach the structures have become part of the large 'water purification machine', form prominent features in the landscape, and serve as locations where the process can be experienced.

Subareas in the recreational area. The recreational area is comprised of subareas which each have an individual character: the Druiplanden, Huygendijk wood and Subplan 4. The Druiplanden, with an urban character, offers space for a pop podium with catering establishment and intensive bank recreation (sand beach, sunbathing areas, car parking, a day camp site with waterski). Subplan 4 forms the transition between

| ECOLOGY PARK 生态公园 | CULTURAL PARK 文化公园 | **COMMUNITY PARK 社区公园** | CENTRAL PARK CBD 公园 |

the urban region of Heerhugowaard-South and the recreational area. The subplan offers 'outdoor' living in surroundings with many trees. Spacious recreational routes cross the area and link the recreational area to the urban region of Heerhugowaard-South. The Huygendijk wood has a sheltered character and offers space for walking, cycling, jogging and roller-skating, etc.

The decor for these activities is comprised of forested areas, open (sunbathing) grassland and nature banks.

设计说明

设计师将在这里引入新的水源景观，将住宅区与娱乐区隔开，保证为市民提供充足的开放空间。娱乐区域主要分为两部分，一个是面向开阔水源和城区的，另一个则是面向外围景观的。卢那公园水系宏大而颇具特色，是为在夏天蓄积并保护大量的水资源而设计的。设计师特别关注水系的水质、可达性以及参与体验的可能性。为此，他们设计了一些构筑物，如循环泵站、天然净水泵、降磷化池、渡桥以及一艘体验漂流的独木舟。

这些构筑物的设计使人们可以近距离地感触水体的净化过程。因此，净水站进水口的位置高于水平面，使人们既可观其形，又可闻其声。而循环泵站巧妙地设置于沙滩入口附近，其屋顶宛若一个俯瞰大湖的公共阳台。在整个场所的景观中，所有构筑物都极具特色，让人们可以从不同角度体验水系。这些构筑物是景观的一部分，共同构成了一座大型"净水器"。

娱乐区域是由多个不同的分区共同组成的，每个分区都有不同的特点，Druiplanden区具有明显的城市特征，有餐饮区和密集的滨水娱乐区，例如沙滩、日光浴区、停车场、日间夏令营等；Subplan 4区是城市区域与娱乐区之间的过渡带，这里树木较多，适合进行户外活动；Huygendijk森林、草地、河岸等为市民的各种活动，包括散步、骑自行车、跑步、滑旱冰等活动提供适当的场地。

| LEISURE AND TOURISM LANDSCAPE 休闲度假景观 | TOURISM PARK 旅游度假公园 | CITY PARK 城市公园 |

| ECOLOGY PARK 生态公园 | CULTURAL PARK 文化公园 | COMMUNITY PARK 社区公园 | CENTRAL PARK CBD 公园 |

LEISURE AND TOURISM LANDSCAPE 休闲度假景观

TOURISM PARK 旅游度假公园 | CITY PARK 城市公园

Keywords 关键词
- City Landmark 城市地标
- Ecological Characteristics 生态特征
- Alterations 改建
- Environment 环境

Location: Toronto, Canada
Client: City of Toronto
Landscape Design: Ken Smith Landscape Architect

项目地点：加拿大多伦多市
客　　户：多伦多市政府
景观设计：Ken Smith Landscape Architect

Village of Yorkville Park
约克维尔公园村

Features 项目亮点

As a landmark, the project well preserved the essence of ecology and the original identity of the connective community.

作为地标性的城市公园，项目将该地区的生态精髓和社区连接组织的原始身份很好地保留了下来。

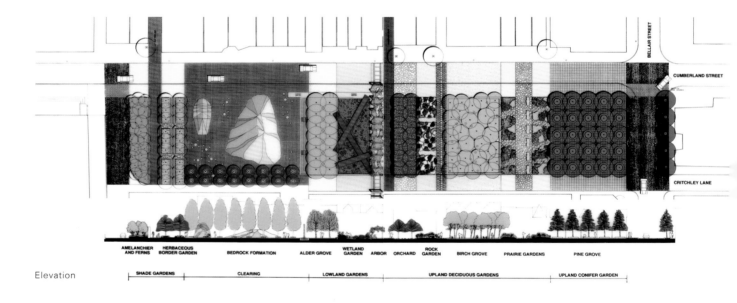

Elevation

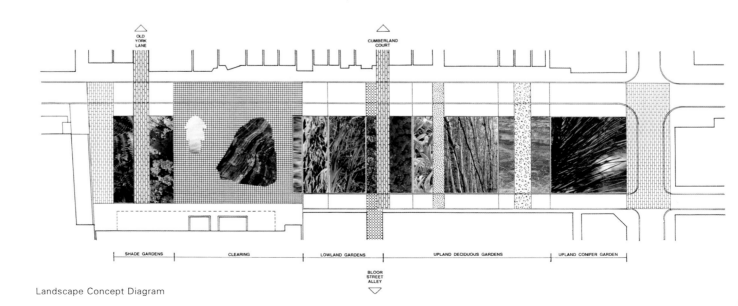

Landscape Concept Diagram

| ECOLOGY PARK 生态公园 | CULTURAL PARK 文化公园 | **COMMUNITY PARK 社区公园** | CENTRAL PARK CBD 公园 |

Overview

The Village of Yorkville Park has become a local landmark. While small in size, the park has played an important role in the revitalization of the neighborhood since its completion in 1994. Recently, the park underwent some restoration work, but its original design integrity as a distillation of regional ecology, along with its role as a neighborhood connection point, remains as strong as ever.

项目概况

约克维尔公园村是多伦多的城市地标。公园于1994年竣工，虽然面积不大，但在复兴周边地区方面扮演着非常重要的角色。目前这个公园正经历一系列改建，但其作为地区生态精髓和社区连接组织的原始身份仍然不可动摇。

LEISURE AND TOURISM LANDSCAPE 休闲度假景观 | TOURISM PARK 旅游度假公园 | CITY PARK 城市公园

| ECOLOGY PARK 生态公园 | CULTURAL PARK 文化公园 | **COMMUNITY PARK 社区公园** | CENTRAL PARK CBD 公园 |

| LEISURE AND TOURISM LANDSCAPE 休闲度假景观 | TOURISM PARK 旅游度假公园 | CITY PARK 城市公园 |

| ECOLOGY PARK 生态公园 | CULTURAL PARK 文化公园 | **COMMUNITY PARK 社区公园** | CENTRAL PARK CBD 公园 |

| LEISURE AND TOURISM LANDSCAPE 休闲度假景观 | TOURISM PARK 旅游度假公园 | CITY PARK 城市公园 |

Keywords 关键词

Native Plants 原生植物

Landscape Lawn 景观草坪

The Cycle Of Rainwater 雨水循环

Green Belt 绿带

Client: ING Real Estate
Location: Tours, Zac des Deux-Lions
Design Team: Agence Nicolas Michelin & Associés, Sempervirens Landscapers Frédéric-Charles AILLET, Raphaël FAVORY, Pierre SARRIEN
Area: 10,000 m²

客　　户：ING Real Estate
项目地点：法国 Zac des Deux-Lions 图尔斯
设计团队：法国 Nicolas Michelin 建筑师事务所，法国 Sempervirens 景观设计事务所，Frédéric-Charles AILLET, Raphaël FAVORY, Pierre SARRIEN
面　　积：10 000 m²

Submersible Garden

法国图尔斯 Submersible 花园

Features 项目亮点

By making the most of the natural slope and ditch to create a series landscape spaces and views along the ditch.

设计很好地利用场地内的自然坡度与沟渠来造景，以河沟为主线串联起一系列的景观空和视线。

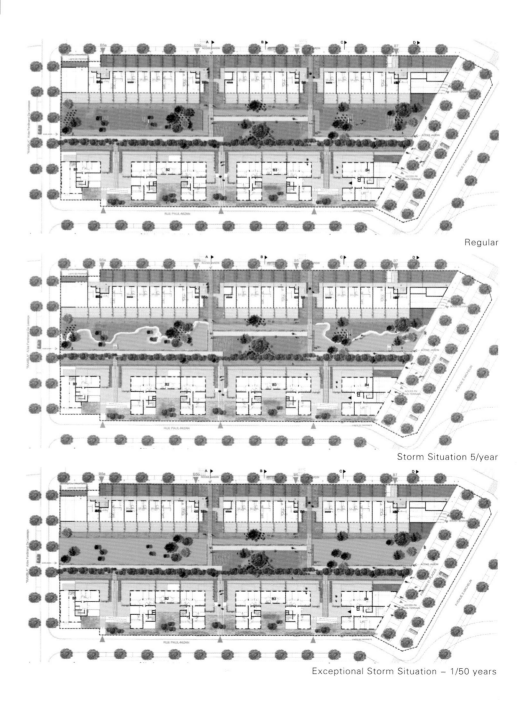

Regular

Storm Situation 5/year

Exceptional Storm Situation – 1/50 years

| ECOLOGY PARK 生态公园 | CULTURAL PARK 文化公园 | **COMMUNITY PARK 社区公园** | CENTRAL PARK CBD公园 |

Design Description

The project was completed by sempervirens Landscape in collaboration with Nicolas Michelin & Associés, a lot of native plants were used during designing process. Stormwater management (50 years flood 800 m³), introduction of flora and fauna native plants.

设计说明

项目由法国Sempervirens景观设计公司与法国Nicolas Michelin 建筑师事务所合作完成，设计过程中大量采用了原生植物。雨水管理（工程使用年限：50年，抗洪量：800 m³）

LEISURE AND TOURISM LANDSCAPE 休闲度假景观

| TOURISM PARK 旅游度假公园 | CITY PARK 城市公园 |

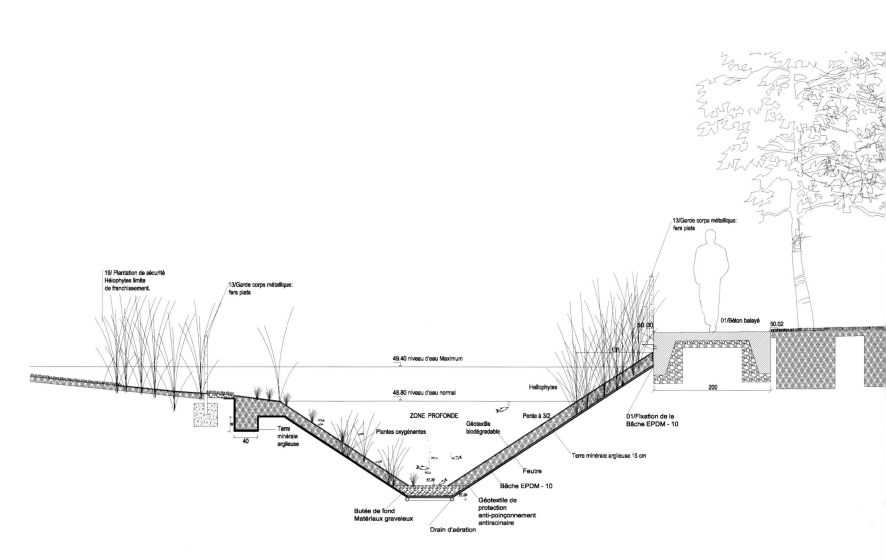

| ECOLOGY PARK 生态公园 | CULTURAL PARK 文化公园 | **COMMUNITY PARK 社区公园** | CENTRAL PARK CBD 公园 |

Urban Plaza
Streetscape
Integrity
Sharing

城市广场
街道空间
整体性
共享性

LEISURE AND TOURISM LANDSCAPE 休闲度假景观

TOURISM PARK 旅游度假公园　　CITY PARK 城市公园

Keywords 关键词

Public Sculpture 公共雕塑
Green Roof 绿色屋顶
Garden Space 花园空间
Sustainability 可持续

Location: St. Louis, USA
Client: City of St. Louis
Landscape Design: Nelson Byrd Woltz Landscape Architects

项目地点：美国圣路易斯
客　　户：圣路易斯市政府
景观设计：Nelson Byrd Woltz Landscape Architects

City Garden
城市花园

Features 项目亮点

By using a variety of sustainable development strategies, including storm water management, native plants, plant maintenance and revitalization of urban areas, the design creates a great public space.

设计运用多种可持续发展策略，包括雨水管理、本土植物、植物健康维护和市区重振，打造了一个深度的公共空间。

Context Plan

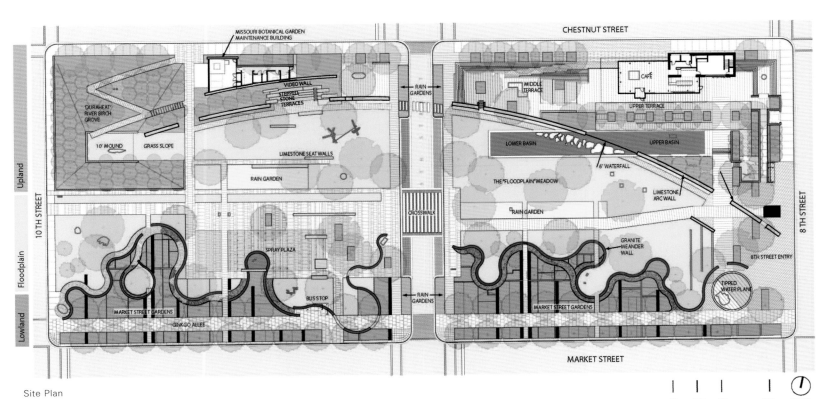

Site Plan

| ECOLOGY PARK 生态公园 | CULTURAL PARK 文化公园 | COMMUNITY PARK 社区公园 | **CENTRAL PARK CBD公园** |

Overview

City garden is a three-acre public sculpture garden created on the Gateway Mall in downtown St. Louis. Sponsored by a private foundation, the garden has played a primary role in reinvigorating the city's center. The design weaves innovative stormwater management strategies with abstractions of local geology, hydrology, and plant communities to create a multi-faceted public space that has become a magnet for locals and tourists alike.

项目概况

城市花园是一个私人基金会赞助的公共雕塑项目，占地3英亩，位于圣路易斯市中心大门廊商场旁。这个花园振兴了城市中心。设计将雨水管理与场地地质、水文、植物群落一并考虑，建立具有深度的公共空间，成为吸引游客和市民的好去处。

Before Construction June 2008

After Construction June 2009

LEISURE AND TOURISM LANDSCAPE 休闲度假景观

| TOURISM PARK 旅游度假公园 | CITY PARK 城市公园 |

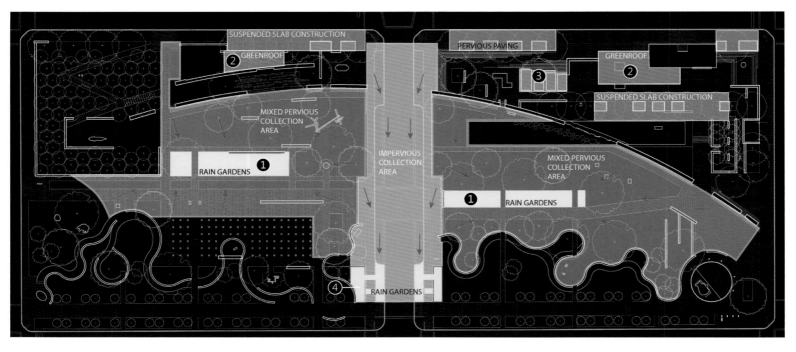

② Greenroof

③ Pervious Paving

④ Steel Grate Over Rain Garden

| ECOLOGY PARK 生态公园 | CULTURAL PARK 文化公园 | COMMUNITY PARK 社区公园 | **CENTRAL PARK CBD公园** |

Design Description

City garden employs multiple sustainable strategies including stormwater management, native planting, promotion of plant health, and reinvigoration of the social/economic health of a declining urban center. Two-thirds of the site's stormwater drainage is managed within the boundaries of the garden. Nearly half of the surface area is permeable. More than 5,000 square feet of rain gardens capture, retain, and infiltrate surface water flow. Three different soil mixes were designed to replace the rubble-filled, over-compacted existing site soils. These soils provide a much healthier substrate for the various plant communities to thrive in. Trees planted within pavement zones are given generous room for trunk and root growth.

设计说明

设计运用多种可持续发展策略，包括雨水管理、本土植物、植物健康维护和市区重振。三分之二的雨水排水区域由花园内部消化，一半的地表为渗水地面。超过 465 m² 的雨水花园截住、保持并过滤地表水流。三种不同类型的土壤混合物代替场地中的现状土壤。这些新土壤有利于植物的生长。同时硬质铺装的设计充分考虑到对植物根系的影响，以确保植物更好地成活。

LEISURE AND TOURISM LANDSCAPE 休闲度假景观

| TOURISM PARK 旅游度假公园 | CITY PARK 城市公园 |

| ECOLOGY PARK 生态公园 | CULTURAL PARK 文化公园 | COMMUNITY PARK 社区公园 | **CENTRAL PARK CBD 公园** |

LEISURE AND TOURISM LANDSCAPE 休闲度假景观

TOURISM PARK 旅游度假公园 CITY PARK 城市公园

Keywords 关键词

Public Space 公共空间
Environment 环境
Restoration 修复重建
Natural Materials 天然材料

Location: New York City, USA
Client: Bryant Park Corporation
Landscape Design: OLIN

项目地点：美国纽约市
客　　户：布莱恩特公园管理委员会
景观设计：OLIN景观公司

Bryant Park
布赖恩特公园

Features 项目亮点

Through small adjustments producing significant results, the design restores the urban park and creates a definitive model of environmental, social, and economic sustainability.

设计过程通过诸多能产生显著成果的小调整来恢复这一城市公园，打造出环境、社会、经济可持续性的绝佳典范。

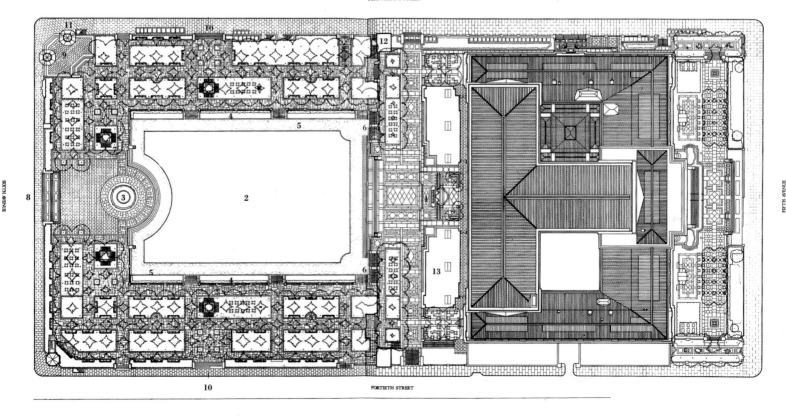

总平面图

| ECOLOGY PARK 生态公园 | CULTURAL PARK 文化公园 | COMMUNITY PARK 社区公园 | **CENTRAL PARK CBD 公园** |

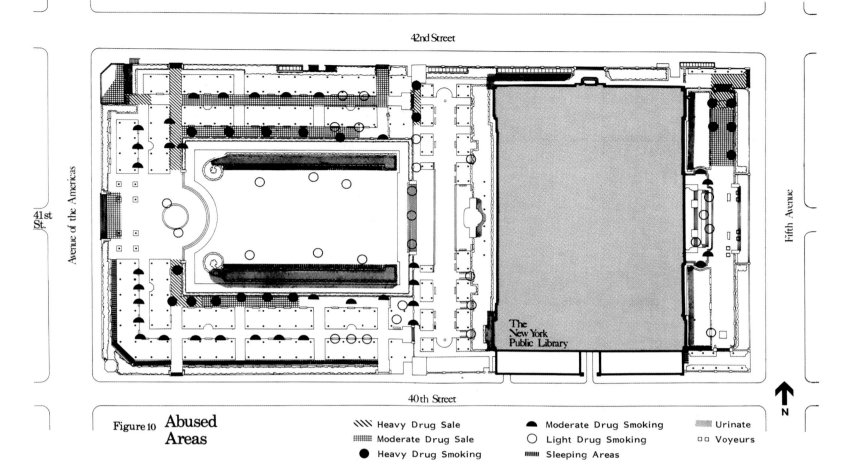

Figure 10 **Abused Areas**

\\\\ Heavy Drug Sale ● Moderate Drug Smoking ▨ Urinate
▦ Moderate Drug Sale ○ Light Drug Smoking ▯▯ Voyeurs
● Heavy Drug Smoking ▥ Sleeping Areas

253

| LEISURE AND TOURISM LANDSCAPE 休闲度假景观 | TOURISM PARK 旅游度假公园 | CITY PARK 城市公园 |

Overview

As a high-profile and cherished public space in one of the world's most important urban centers, the enormously successful restoration of Bryant Park was pivotal for New York City and serves as the definitive model of urban park restoration and environmental, social and economic sustainability.

项目概况

作为一个世界上最重要城市的宝贵的中心绿地，布赖恩特公园恢复成为公共空间是巨大的成功，这不光对纽约市很关键，也是一个城市公园恢复以及环境、社会、经济可持续性的最佳典范。

| ECOLOGY PARK 生态公园 | CULTURAL PARK 文化公园 | COMMUNITY PARK 社区公园 | **CENTRAL PARK CBD 公园** |

Design Description

The design included numerous small changes that would yield significant results. The landscape architect modified and added entrances, ramps, stairs, and pavements while cutting through walks and railing to configure free circulation. The design also included concessions, public restrooms, and entertainment programming as critical components of a successful restoration. During the design process, stone paving from the 1934 scheme was salvaged for reuse and gravel walks were introduced at the heart of the park. The derelict hedges in the central area that had proved to be a serious maintenance and safety problem were removed. The lawn was enlarged and two 300-foot-long borders containing herbaceous perennials and evergreens were placed against the walls and railings where they could be seen and enjoyed without forming a physical or visual barrier. The restoration's long-term environmental sustainability was insured by the landscape architect's commitment to incorporating durable and natural materials throughout the park.

设计说明

设计包含了很多能产生显著成果的小调整。景观设计师在修改补充入口、坡道、楼梯和人行路的同时，还减少栅栏的配置以达到更自由地流通。设计还包括在主要区域设置公共休息室和植入娱乐。在设计过程中，1934 年的石子路被重新翻新并在公园的中心区域设置相关介绍。中部需要严重维修和缺乏安全的设置则被拆除。把草坪扩大，两侧 300 英尺（91.44 m）长的边界上种栽多年生草本植物和常青草，这一处理方法并没有形成一个物理或视觉障碍。景观建筑师承诺恢复公园长期可持续性的环境，并使用了天然材料。

| LEISURE AND TOURISM LANDSCAPE 休闲度假景观 | TOURISM PARK 旅游度假公园 | CITY PARK 城市公园 |

| ECOLOGY PARK 生态公园 | CULTURAL PARK 文化公园 | COMMUNITY PARK 社区公园 | **CENTRAL PARK CBD 公园** |